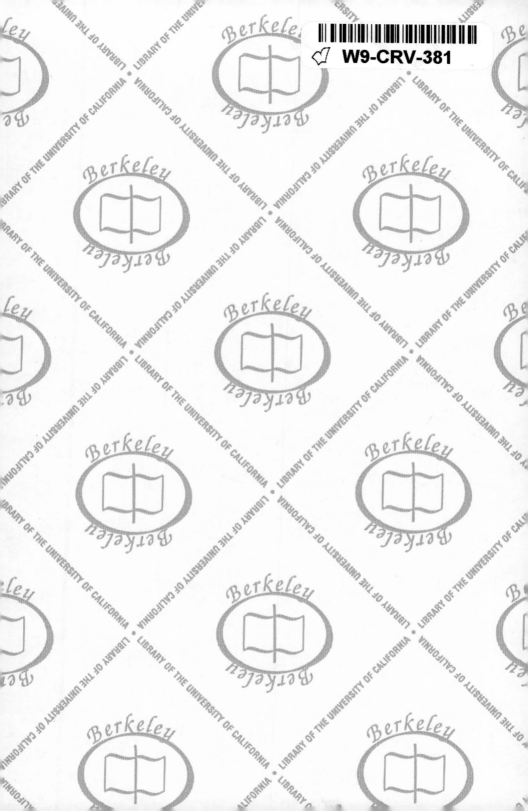

Chemistry In Alternative Reaction Media

Chemistry In Alternative Reaction Media

Dave J. Adams
University of Leicester, UK

Paul J. Dyson
École Polytechnique Fédérale de Lausanne, Switzerland

and

Stewart J. Tavener
University of York, UK

WILEY

Other Wiley Editorial Offices

John Wiley & Sons Inc., 111 River Street, Hoboken, NJ 07030, USA

Jossey-Bass, 989 Market Street, San Francisco, CA 94103-1741, USA

Wiley-VCH Verlag GmbH, Boschstr. 12, D-69469 Weinheim, Germany

John Wiley & Sons Australia Ltd, 33 Park Road, Milton, Queensland 4064, Australia

John Wiley & Sons (Asia) Pte Ltd, 2 Clementi Loop #02-01, Jin Xing Distripark, Singapore 129809

John Wiley & Sons Canada Ltd, 22 Worcester Road, Etobicoke, Ontario, Canada M9W 1L1

Wiley also publishes its books in a variety of electronic formats. Some content that appears
in print may not be available in electronic books.

Library of Congress Cataloging-in-Publication Data

Chemistry in alternative reaction media / Dave Adams, Paul Dyson, and Stewart Tavener.
    p. cm.
  Includes bibliographical references and index.
    ISBN (invalid) 0-471-49849-1 (alk. paper) – ISBN 0-471-49849-1 (paper : alk. paper)
    1. Solvents. 2. Solvation. 3. Chemical reactions. 4. Chemical kinetics. I. Adams, Dave
(Dave J.) II. Dyson, Paul J. III. Tavener, Stewart.

QD544.C44 2004
541.3′9 – dc21

                                                                          2003047967

British Library Cataloguing in Publication Data

A catalogue record for this book is available from the British Library

ISBN 0-471-49848-3 (Cloth)
ISBN 0-471-49849-1 (Paper)

Typeset in 10/12pt Times by Laserwords Private Limited, Chennai, India
Printed and bound in Great Britain by Biddles Ltd, Guildford, Surrey
This book is printed on acid-free paper responsibly manufactured from sustainable forestry
in which at least two trees are planted for each one used for paper production.

CONTENTS

Preface

Sipping a cup of decaffeinated coffee the reader may wonder on the somewhat unusual classification of solvents as 'alternative': alternatives to what? And why would we need alternative media for doing chemistry or for any other purpose? These may be the first questions of those who are just starting to discover the exciting new developments on using solvents other than volatile and often toxic organics for synthesis and especially for catalytic synthetic reactions. Yes, indeed, the vast majority of synthetically useful reactions do not take place between isolated entities in the gas phase; the science of chemistry has been developed and still mostly practiced in solutions where the reactants of the chemical transformation are molecularly dispersed by virtue of their interaction with the molecules making the majority of the liquid phase, i.e. the solvent. The thermodynamics of solutions is one of the oldest and most meticulously developed field of physical chemistry and serves as a fundament for chemical engineering. Solvents for a reaction are chosen according to their ability to dissolve the important ingredients of the reaction: solubility of the reactants, products and often the intermediates should carefully be considered. Traditionally, and logically from solubility aspects, the most important syntheses are run in organic solvents, but it is here, where several problems accumulated over the years. First and foremost, more and more organic solvents are blacklisted due to their damaging effect on human health or on the environment in general. Second, the synthesis of fine chemicals requires catalysts of highly sophisticated architecture and this is synonymous to those being extremely expensive. Such precious catalysts simply cannot be allowed to decompose during the workup phase of synthesis, often involving distillation.

From these considerations several new approaches emerged to reduce the risks and disadvantages of using traditional organic solvents by replacing them with so called *alternative* ones, the list of which include water, fluorous solvents, supercritical fluids and ionic liquids. Note, that *alternative* does not necessarily mean *new* or newly discovered; in fact, most of these solvents have been around for several decades (not even mentioning water). What is really new is their use as alternatives to common organic solvents. All these liquids can be applied as the sole solvent in a given reaction, but most often biphasic or multiphasic reaction media are designed on the basis of the limited solubility of these liquids in each other. One should even add, that a proper combination of traditional organic solvents may result in a biphase. By now probably all the possible permutations of the mentioned five general solvent classes have been demonstrated applicable for multiphase synthesis, the best known examples being the

Shell higher olefin process (organic/organic) and the Ruhrchemie-Rhône Poulenc propene hydroformylation process (aqueous/organic). The diversity of the applications may confuse the newcomer but it is not easy to comprehend even by the more experienced. A guide to this field may help a lot, and this is why the book of Adams, Dyson and Tavener is most welcome.

Of course, there are numerous reviews and several edited books on certain aspects of the use of a given solvent class in organic (and more rarely, inorganic) synthesis. But this book represents the first approach to describe the properties, physical chemical characterization and synthetic use of water, fluorous solvents, supercritical fluids and ionic liquids from a unified view. Numerous tables and graphs help the reader not only to learn about the properties of these alternative solvents (not widely available elsewhere, and certainly not in the same publication) but also to discover the relation between the chemical structure and solvent ability. Perhaps the most attractive feature of the book is the well balanced combination of elementary physical chemistry (dissolution, solvent properties and characterization, polarity scales, solvent effects on kinetics, and the like) with the most advanced use of these principles to characterize such hard-to-handle solvents as supercritical fluids and ultrapure ionic liquids. In this way the authors succeed in bridging the gap between the often unusual features of the so-called alternative reaction media and their sound interpretation by chemical thermodynamics and kinetics. To see a substance, composed entirely of ions, flowing like water at room temperature may be a shocking experience, but this is not more shocking than to feel the sharp *fall* of temperature in a vigorous endothermic reaction made possible by a large increase in entropy (such as, for instance, the one between thionyl chloride and hydrated cobalt(II) chloride). Although in our everyday life we are not *used to* such phenomena they still surround us and it is just more than appropriate to make use of the unusual properties of *alternative solvents* in catalysis and synthesis, beyond the extraction of caffeine from the beans with supercritical carbon dioxide.

This book of Adams, Dyson and Tavener does not intend to give a last minute overview of all the applications using alternative solvents for chemical purposes; taking the fast pace with which research in this area moves this would be a vain attempt anyway. What they do, instead, is the presentation of the main concepts and the description of the most important and perhaps most representative examples of biphasic catalytic processes; these are supplemented by carefully chosen literature references. Moreover, experimental techniques to study such reactions are also described that will prove most useful in such unusual cases as, the investigations applying supercritical fluids. The outspoken intention of the authors is to give a guide, based on solid thermodynamic and kinetic basis, for those who wish to be familiar with the properties and with the fascinating possibilities alternative solvents offer to devise more efficient and less hazardous (green) chemical processes. I am convinced, that the prospective readers, from university students to industrial experts, will all find this book timely, clearly written and useful in their work.

After all: we all need more efficient and less hazardous chemical processes. And for that reason we all need the use of alternative solvents.

July 11, 2003.

Ferenc Joó
University of Debrecen

Abbreviations and Acronyms

α	Kamlet–Taft hydrogen bond donor parameter
AIBN	2,2′-azobis(isobutyronitrile)
AN	electron pair acceptor number
β	Kamlet–Taft hydrogen bond acceptor parameter
Bp	boiling point
bipy	2,2′-bipyridine
c	cohesive pressure (cohesive energy density)
CED	cohesive energy density (cohesive pressure)
CFC	chlorofluorocarbon
cod	cyclooctadiene
δ	chemical shift (in NMR)
δ	Hildebrandt's solubility parameter
dba	dibenzylideneacetone
DMF	N,N-dimethylformamide
DMSO	dimethyl sulfoxide
DN	electron pair donor number
DN^N	normalized electron pair donor number
dppe	bis(diphenylphosphino)ethane
dppf	1,1′-bis(diphenylphosphino)ferrocene
ε_0	permittivity of a vacuum
ε_r	dielectric constant or relative permittivity
ee	enantiomeric excess
$E_T(30)$	energy of transition for Reichardt's betaine dye
E_T^N	normalized energy of transition for Reichardt's betaine dye
FC-72	n-perfluorohexane
FRPSG	fluorous reverse phase silica gel
HBA	hydrogen bond acceptor
HBD	hydrogen bond donor
HOMO	highest occupied molecular orbital
HPLC	high pressure liquid chromatography
k, k_1, k_2, k_{obs}	rate constants
IL	ionic liquid
IR	infrared
K	equilibrium constant
l	length
LDPE	low density polyethylene

LUMO	lowest unoccupied molecular orbital
μ	dipole moment
m	mass
Mp	melting point
MS	mass spectrometry
NAILs	nonaqueous ionic liquids
nbd	norbornadiene
NCW	near-critical water
NMR	nuclear magnetic resonance
π^*	Kamlet–Taft general polarity parameter
π	internal pressure
P	partition coefficient
P	pressure
P_c	critical pressure
PP2	perfluoromethylcyclohexane
PP3	perfluorodimethylcyclohexane
PTC	phase transfer catalysis/catalyst
Q, Q$^+$	quaternary ammonium or phosphonium (or other cation in phase transfer reactions)
ρ_c	density at critical point
R$_f$	perfluoroalkyl group
RMM	relative molecular mass
sc	supercritical
SCF	supercritical fluid
T_c	critical temperature
TEMPO	2,2,6,6-tetramethylpiperidinyl-1-oxy
THF	tetrahydrofuran
TMS	tetramethylsilane
tppms	monosulfonated triphenylphosphine
tppts	trisulfonated triphenylphosphine
UV-vis	ultraviolet – visible spectroscopy
VOCs	volatile organic compounds
z	charge

IONIC LIQUID CATIONS

[mmim]	1,3-dimethylimidazolium
[emim]	1-ethyl-3-methylimidazolium
[bmim]	1-butyl-3-methylimidazolium
[hmim]	1-hexyl-3-methylimidazolium
[omim]	1-octyl-3-methylimidazolium
[dmim]	1-decyl-3-methylimidazolium
[bdmim]	1-butyl-2,3-dimethylimidazolium

1 Chemistry in Alternative Reaction Media

Chemical reactions do not occur in isolation, but within an environment that is dictated by the surrounding molecules, atoms and ions. This environment can be called the *medium*, and may consist of other reactant molecules in the gas phase, or neighbours within a crystal lattice. In many cases, however, a *solvent* of some kind is used as the reaction medium, and the reactants are *solutes*. A solvent and a solute may be defined as follows:

Two compounds which mix together to give a single, homogeneous liquid phase.

In general, the compound present in greatest quantity is the solvent and the other is the solute. Although all compounds may behave as solvents, it is only those that are liquid at room temperature which are usually classed as such. Solvents are widely used in all aspects of chemistry: in synthesis, chromatographic separation, dilution, extraction, purification, analysis and spectroscopy, crystal growth and cleaning [1, 2]. Solvents may also be reactants; reaction of a solvent is known as *solvolysis*. Since any compound is a solvent, at least in principle, then the ability to select the right one for a particular task is essential. A solvent for a particular application might be selected on the following criteria:

1 The effect that the solvent has on the chemical reaction's products, mechanism, rate or equilibrium.
2 The stability of substrates, products and (often delicate) catalysts, transition states and intermediates, in the solvent.
3 Suitable liquid temperature range for useful reaction rates.
4 Sufficient solvent volatility for removal from the product by evaporation or distillation.
5 Cost, which is a particularly important consideration when scaling up for industrial applications.

This chapter outlines the major uses and characteristics of solvents, and discusses some of the problems associated with their use. The remainder of the book describes some of the alternative media in which chemistry may be conducted, and it is hoped that these alternatives will give advantages over conventional

Chemistry in Alternative Reaction Media   D. Adams, P. Dyson and S. Tavener
© 2004 John Wiley & Sons, Ltd   ISBNs: 0-471-49848-3 (Cloth); 0-471-49849-1 (Paper)

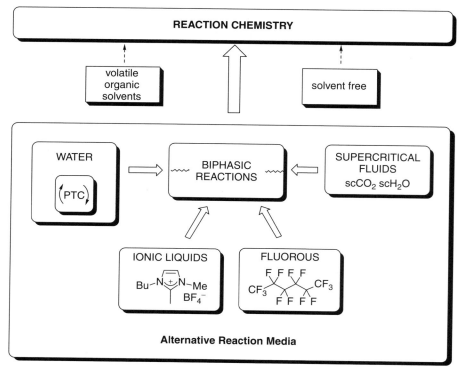

Figure 1.1   Summary of alternative reaction media

volatile organic solvents in terms of improved ease of separation, efficiency and yield. They may even open doors to new reaction chemistry. These alternatives may be used by themselves or in the multiphasic systems described in Chapter 2. Figure 1.1 outlines the solvent systems covered in this book.

1.1   ECONOMIC AND POLITICAL CONSIDERATIONS

Whilst this book focuses on the chemical uses of solvents, it should not be forgotten that, of an estimated 60 million tonnes of synthetic solvents used each year, a large proportion is used for non-chemical applications. Vapour degreasing, dry cleaning and immersion-cleaning of mechanical parts are amongst the largest of these. Figure 1.2 shows some of the major uses of solvents.

Despite their evident utility, the use of solvents in chemical processes must be scrutinized from environmental and economic points of view because solvent use is inherently wasteful. In a chemical process, a solvent is usually added to reactants to facilitate reaction, and is later removed from the chemical product prior to disposal or, preferably, recycling and reuse. Removal of residual solvent from

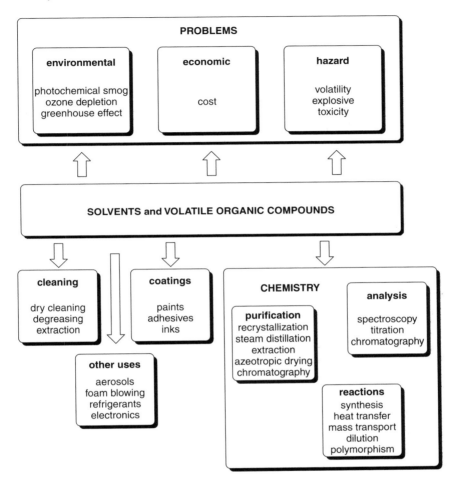

Figure 1.2 Summary of the major uses and problems associated with solvents

a chemical product is frequently achieved either by evaporation or distillation, and for this reason most popular solvents are highly volatile. This volatility has led to some major public concerns about solvent use, some of which appear in the 'problems' box in Figure 1.2. Leaks and spillage of volatile solvents inevitably lead to evaporation into the atmosphere. Atmospheric pollution has been one of the major global environmental issues of the late 20th and early 21st century, and emissions of some categories of volatile organic solvents have been implicated in the depletion of the Earth's ozone layer and in the greenhouse effect. On a local level, exposure to volatile organic compounds (VOCs) in the workplace may lead to dizziness, nausea and other, longer term effects including respiratory problems and cancer. In 1987, the Montreal Protocol set a timescale for the reduction and eventual phasing out of greenhouse gases and ozone depleting compounds. In particular, the use of chlorofluorocarbons (CFCs), bromofluorochlorocarbons (halons) and carbon tetrachloride as solvents, refrigerants, fire extinguishers and aerosol propellants is now a thing of the past [3]. In the mid-1990s, the widely used cleaning agent and degreaser 1,1,1-trichloroethane was removed from use because of its ozone depleting effects. In 1991 the Geneva Protocol set a framework for the reduction of VOC emissions, and the Kyoto Protocol of 1997 set targets for the reduction of six more classes of emission believed responsible for climate change: CO_2, nitrogen oxides (NO_x), hydrofluorocarbons, volatile perfluorocarbons, methane and sulfur hexafluoride. More recent legislation in the United States and in Europe has centred around the emissions of VOCs. Other compounds have come under scrutiny because of their toxicity and carcinogenicity. Government bodies and environmental agencies are implementing these legislative measures with a carrot-and-stick approach. Funding initiatives to aid and promote the development of cleaner chemical technologies are balanced by fines and penalties to punish those responsible for emission of pollutants into the environment [4]. The principal of 'the polluter pays' is now in place both in Europe and in the USA [5]. Unfortunately, many of the compounds involved are exactly those that have desirable properties as solvents. The legislation process is ongoing and the chemical industry is accordingly looking for both short term replacements for controlled solvent substances, and long term strategies that will make manufacturing processes conform with future controls. It seems highly likely that, in the near future, all emissions of organic compounds to the environment will be strictly controlled, and the more cynical might classify organic solvents as the 'banned' and the 'soon-to-be-banned'. Table 1.1 shows a generalized guide to the acceptability of solvent types, although this should be viewed with caution. For example, methanol, an oxygenated solvent widely considered to be acceptable, is a suspected carcinogen.

The industries involved in the manufacture of solvents and formulations which use them have been hit hard by antipollution legislation. Many solvent users have responded to these regulations by reformulating products to reduce their content of volatile organic components, or even eliminate them completely [6]. These

Table 1.1 General guide to the acceptability of solvents

Most acceptable	None (rarely possible)
	Water
	Oxygenated (e.g. alcohols, ethers, ketones and esters)
	Aliphatic hydrocarbons (e.g. cyclohexane, dodecane)
	Aromatic hydrocarbons (e.g. xylene, mesitylene)
	Dipolar aprotic (e.g. DMSO, DMF, NMP)
	Chlorinated solvents (e.g. dichloromethane)
	Ozone depleters (e.g. CFCs, 1,1,1-trichloroethane)
Least acceptable	Toxic and carcinogenic solvents (CCl_4, benzene)

new formulations include aqueous emulsions, high solids paints, and radiation-curable or powder based coatings. In addition, much more consideration is now given to the recycling of solvents within a manufacturing plant. Current forecasts predict a continuation of the recent trends seen in the USA and shown in Table 1.2: a decline in the demand for aromatic, aliphatic hydrocarbon and chlorinated solvents, and an increased demand for oxygenated alternatives [7]. Oxygenated solvents are more acceptable because they are believed to break down in the environment much more quickly than hydrocarbon and halogenated solvents. Sales of chlorinated solvents in Western Europe fell by 17 % between 1996 and 2001 (Figure 1.3), although this statistic discounts recycling and so

Table 1.2 Trends in solvent demand in the USA (10^3 tonnes per annum)

Solvent class	1987	2001
Hydrocarbons	3244	1372
Ethers	1239	1645
Alcohols	1124	1525
Ketones	506	480
Chlorinated	805	330
Esters	216	355
Other solvents	180	227
Total	7244	5934

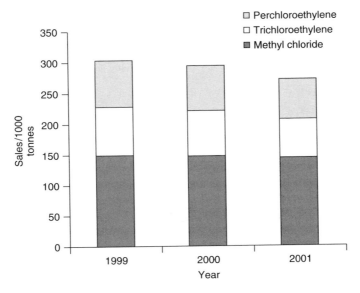

Figure 1.3 Sales of chlorinated solvents in Western Europe 1996–2001 [10]

may represent improvements in efficiency as well as reduction in use [8]. These figures are based on solvent sales, and, when recycling and reuse are taken into account, it is estimated that the actual quantity of solvents used may be as high as 60 million tonnes per year for the USA alone [9].

1.2 WHY DO THINGS DISSOLVE?

A substance will dissolve, quite simply, if it is energetically favourable for it to do so. If the sum of the energies required to break apart those forces holding the potential solute together *and* to separate the solvent molecules from one another is outweighed by the energy released on solvation, then the substance will dissolve. However, dissolution is a kinetic as well as a thermodynamic process, and solutes that dissolve slowly may be accelerated greatly by the employment of heat or ultrasound.

To understand the dissolution of ionic solids in water, lattice energies must be considered. The lattice enthalpy, ΔH_l, of a crystalline ionic solid is defined as the energy released when one mole of solid is formed from its constituent ions in the gas phase. The hydration enthalpy, ΔH_h, of an ion is the energy released when one mole of the gas phase ion is dissolved in water. Comparison of the two values allows one to determine the enthalpy of solution, ΔH_s, and whether an ionic solid will dissolve endothermically or exothermically. Figure 1.4 shows a comparison of ΔH_l and ΔH_h, demonstrating that AgF dissolves exothermically.

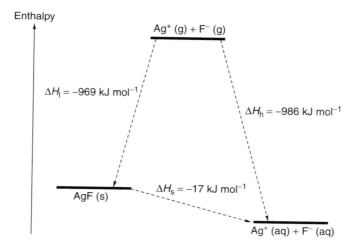

Figure 1.4 Thermodynamic cycle comparing the lattice enthalpy, ΔH_l, and the enthalpy of hydration, ΔH_h, for AgF. The enthalpy of solvation, ΔH_s, is equal to the difference between ΔH_l and ΔH_h

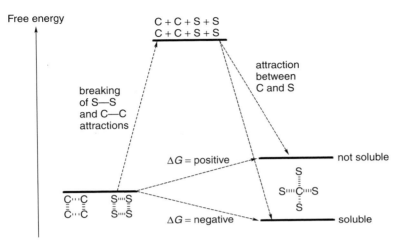

Figure 1.5   Thermodynamic cycle for the dissolution of compound C in solvent S. For disso-lution to occur, the attractive forces between solute and solvent must outweigh the free energy required to separate solvent–solvent and solute–solute attractive forces

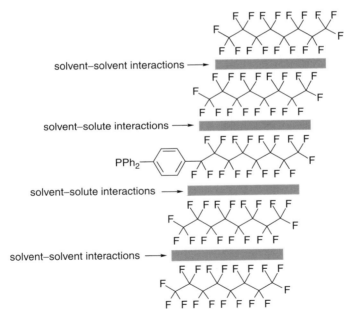

Figure 1.6   'Like dissolves like': perfluoroalkyl ponytails make phosphines more soluble in a fluorous solvent. These phosphines are suitable ligands for metal catalysts, and will therefore aid the solubility of these catalysts in fluorous solvents

For an ionic compound to be soluble in water, the free energy of solution, ΔG_s ($\Delta G = \Delta H - T \Delta S$), must be negative. In general, ΔH_s values are usually also negative, but there are exceptions. For example, NaCl has a small, positive value for ΔH_s (about $+2 \, kJ \, mol^{-1}$). It dissolves endothermically, and dissolution of salt in water causes a cooling of the solution. Here, the entropy terms must be playing a significant role.

Dissolution of non-ionic compounds may be considered in a similar manner. For a compound C to dissolve in solvent S, the free energy of mixing must be negative; it needs to outweigh the breaking of the attractive forces C–C and S–S, as shown in Figure 1.5. This applies to the mixing of liquids as well as the dissolution of solids and is the basis of the hydrophobic effect, which is discussed in Chapter 5 in the context of the use of water as a solvent for organic chemistry.

The general rule for solvation is that 'like dissolves like'. A compound which has a particular functional group attached will often dissolve well in a solvent that contains that functional group. Perfluorinated aliphatic solvents are good examples to mention here as they are being explored as solvents for multiphasic catalysis (see Chapter 3). In order to dissolve metal catalysts in these solvents, lengthy perfluorinated groups are attached to the ligands surrounding the catalyst. These groups should enable the solute–solvent interactions to be as similar as possible to solvent–solvent attractions, as shown in Figure 1.6.

1.3 SOLVENT PROPERTIES AND SOLVENT CLASSIFICATION

The diversity of solvents makes classification very complex and many different ways of classifying solvents are used. Solvents may be broadly classed according to their chemical type, i.e. aqueous, molecular-inorganic (composed of covalently bonded molecules, e.g. NH_3), molecular-organic, ionic (made up of cations and anions) or atomic (noble gases or metallic) liquids. Molecular organic solvents may be further divided according to their chemical composition – aliphatic, aromatic, alcohol or other functional group. Solvents are often classed according to their physical properties. Key properties include melting and boiling points, viscosity, density, dipole moment, dielectric constant, specific conductivity and cohesive pressure (see Table 1.3). The physical properties that are considered most important depend upon the application. For example, in a synthesis that involves conducting a reaction at elevated temperature, then boiling point may be the most important constant. However, knowing the dielectric constant of the solvent is also essential if microwave heating is to be used [11].

1.3.1 Density

Density (the mass of a compound per unit volume) is an important factor to consider for the separation of immiscible liquids. Two phases should have sufficient

Table 1.3 Properties used to classify solvents

Molecular physical property	Dipole moment
Bulk physical properties	Cohesive pressure
	Dielectric constant
	Refractive index
	Melting point and boiling point
Chemical properties	Donor numbers
	Acceptor numbers
Solvatochromic properties	E_T^N, α, β and π^*

density difference to ensure efficient separation, otherwise an emulsion may be formed. The relatively high density of bromoform ($CHBr_3$, $\rho = 2.9\,g\,cm^{-3}$) and other halogenated solvents is utilized in the mining industry to separate crushed heavy ore-rich minerals from lighter gangue[1] material. Densities of liquids range from about $14\,g\,cm^{-3}$ for liquids metals (Hg), through $1-3\,g\,cm^{-3}$ for halogenated solvents, to under $1\,g\,cm^{-3}$ for hydrocarbons.

1.3.2   Mass Transport

Solvents can increase reaction rates by dispersing reactant molecules and increasing the collision frequency (Figure 1.7a). In solution, all of the solutes are potential reactants. Reactions between solids, however, tend to be much slower than reactions in liquids as there is only a small amount of contact between the solid reactants. Even fine powders will have a relatively small surface area-to-mass ratio, so the bulk majority of the reactant is not in the right place to react (Figure 1.7b).

Exceptional solid–solid reactions with high rates have been observed, but many of these are condensation reactions which eliminate water [12] and it is likely that the reaction takes place in a thin, aqueous layer at the boundary between the solid surfaces, as shown schematically in Figure 1.7c. Other examples are those which produce a liquid product. For example, dimethylimidazolium tetrachloroaluminate, an ionic liquid, may be prepared simply by mixing together dimethylimidazolium chloride with aluminium trichloride [13]. The powdered reactants will collapse together to form a liquid product. This reaction is described further in Chapter 4, in the context of preparing ionic liquids for use as reaction solvents. Despite these inherent drawbacks, there is sometimes enough molecular motion within solids to give a useful rate of reaction, and a wide range of solid–solid reactions have been reported, including oxidations, reductions, additions, eliminations, substitutions and polymerizations [14].

---

[1] *Gangue* is a term used in mining and geology to describe the rock in which a vein of mineral ore is imbedded.

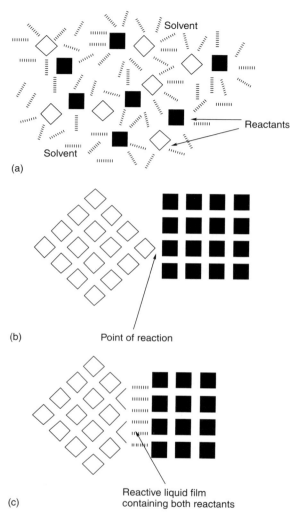

Figure 1.7 (a) In the liquid phase, molecules of both compounds are dispersed and all molecules are potentially reactive; (b) when two solids react, reaction can only occur at the point where the two crystallites are in direct contact; (c) reaction between solids may form a thin liquid layer which increases the rate of reaction

1.3.3   Boiling Point, Melting Point and Volatility

Melting points and boiling points are related to the strength of the intermolecular forces between solvent molecules, and to the molecular weight of the solvent. Dispersive forces, hydrogen bonding and permanent dipole moments all contribute. Typically, for molecules of similar mass, nonpolar compounds which

Table 1.4 Melting points and boiling points of four illustrative compounds

Solvent	RMM	Mp (°C)	Bp (°C)	Intermolecular forces
Propane	44	−188	−42	Dispersive
Acetonitrile	41	−44	82	Permanent dipole
Ethanol	46	−114	78	H-bonding
LiCl	42.4	605	>1300	Electrostatic/ionic

Source: *CRC Handbook of Chemistry and Physics*, 80th Edition, CRC Press, Florida, 2000.

display only dispersive interactions have lower melting and boiling points than those which exhibit permanent dipoles, electrostatic forces, and hydrogen bonding (Table 1.4).

1.3.4 Solvents as Heat-Transfer Media

Solvents play an important role as a heat transfer medium. They carry away heat liberated by an exothermic reaction, or can supply the thermal energy required to initiate an endothermic one. Diffusion and mobility of the solvent reduce the extent of thermal gradients within a reactor, allowing a reaction to proceed in a smooth and even fashion. The ability of a solvent to transfer heat in a reactor is illustrated by the familiar, yet ingenious, concept of reflux (Figure 1.8), in which the reaction temperature is held constant by allowing a portion of the solvent to boil away and condense on a cooled surface, before being returned to the main reservoir of the reactor. This prevents an exothermic reaction from 'running away' and overheating. Solvents with a large degree of intermolecular attraction, and in particular H-bonding solvents, have high heat capacities.

1.3.5 Cohesive Pressure, Internal Pressure, and Solubility Parameter

The *cohesive pressure* (*c*) of a solvent, otherwise known as *cohesive energy density* (CED), is a measure of the attractive forces acting in a liquid, including dispersive, dipolar and H-bonding contributions, and is related to the energy of vaporization and the molar volume (Equation 1.1):

$$c = \frac{\Delta U_{vap}}{V_{molar}} \tag{1.1}$$

Like other measures of pressure, c has units of MPa. In theory, a liquid will break all solvent–solvent interactions on vaporization, and so c is a measure of the sum of all the attractive intermolecular forces acting in that liquid. Hydrogen-bonding and dipolar solvents therefore have high c values. Water has a large value for c, and fluorocarbons very low values (Table 1.5).

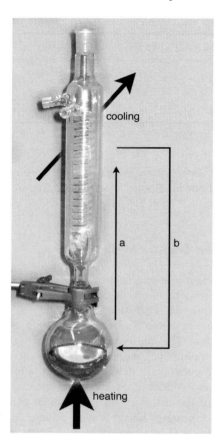

Figure 1.8 Heat transfer via reflux. Solvent evaporates on heating (a) and condenses in a cooler part of the reactor. The cooled liquid is returned to the reservoir (b) and the net effect is the removal of heat from the reservoir

The square route of the cohesive pressure is termed *Hildebrand's solubility parameter* (δ). Hildebrand observed that two liquids are miscible if the difference in δ is less than 3.4 units, and this is a useful rule of thumb. However, it is worth mentioning that the inverse of this statement is not always correct, and that some solvents with differences larger than 3.4 are miscible. For example, water and ethanol have values for δ of 47.9 and 26.0 MPa$^{0.5}$, respectively, but are miscible in all proportions. The values in the table are measured at 25 °C. In general, liquids become more miscible with one another as temperature increases, because the intermolecular forces are disrupted by vibrational motion, reducing the strength of the solvent–solvent interactions. Some solvents that are immiscible at room temperature may become miscible at higher temperature, a phenomenon used advantageously in multiphasic reactions.

Table 1.5 Cohesive pressures (c), Hildebrand's solubility parameter (δ), and internal pressures (π) for a range of representative solvents [1, 2]

Solvent	c (MPa)	δ (MPa$^{0.5}$)	π (MPa)
Water	2302	47.9	151
Methanol	887	29.6	285
Ethanol	703	26.0	291
Acetonitrile	590	24.3	379
Dichloromethane	414	20.3	408
Acetone	398	20.2	337
Chloroform	362	19.0	370
Benzene	357	18.8	379
Ethyl acetate	347	18.6	354
Toluene	337	18.4	379
Cyclohexane	285	17.6	326
Diethyl ether	251	15.8	264
n-Hexane	225	14.9	239
Perfluoroheptane	136	11.9	220

The *internal pressure* (π) of a solvent represents the energy change that the liquid must undergo during a very small increase in volume at constant temperature. This small expansion does not disrupt the H-bonding network of the solvent, and so it is a good indication of the dipolar and dispersive attractions. The data in Table 1.5 show a trend of increasing π value as polarity increases, except for strongly H-bonding liquids such as water and alcohols, which indicates just how strongly the H-bonds contribute to the properties of these solvents. Internal pressure can have an effect on the rates of chemical reactions that display a significant change in the volume, $\Delta V^{\ddagger}$ (*volume of activation*), of the transition state. This is discussed further in Chapter 7.

1.4 SOLVENT POLARITY

It is perhaps obvious that consideration of physical properties such as melting point, boiling point and viscosity of a solvent are essential when choosing a solvent for a particular application. For a chemical reaction, it is also vital that one has some understanding of how well the substrates, reagents and products will dissolve. This is governed by a number of factors, which together make up the character of the solvent. This general character of the solvent is frequently termed *polarity* but, unfortunately, the concept of solvent polarity is not a simple one. For example, methanol is clearly more polar than cyclohexane, but what about dichloromethane, diethyl ether and benzene? We know empirically that these solvents behave differently, but which is most polar? The answer depends on which property we look at: the dielectric constants follow the order dichloromethane > diethyl ether > benzene, but the order of ability to accept an electron pair is dichloromethane > benzene > diethyl ether. Solvent polarity might best be defined as the solvation power of a solvent, and depends on the interplay of electrostatic, inductive, dispersive, charge-transfer and H-bonding forces [15]. Despite the problems of quantifying solvent polarity, numerous methods have been devised to assess polarity based on various physical and chemical properties. These include dielectric constant, electron pair acceptor and donor ability, and the ability to stabilize charge separation in an indicator dye.

1.4.1 Dipole Moment and Dispersive Forces

Any compound with a nonsymmetrical distribution of charge or electron density will possess a permanent *dipole moment*, μ, whereas a molecule with a centre of symmetry will have no permanent dipole moment. Dipole moment is proportional to the magnitude of the separated charges, z, and also the distance between those charges, l.

$$\mu = zl \tag{1.2}$$

Dipolar molecules will form localized structures in a bulk solvent by orientation of these permanent dipoles, and Figure 1.9 shows two different ways in which two molecules with permanent dipoles may align with one another. With several molecules and three dimensions, a wide range of localized structures are possible. Dipole moments have units of coulomb metres (C m) or Debyes (D, 1 D = 3.34 × 10^{-30} C m).

As well as these permanent dipole moments, random motion of electron density in a molecule leads to a tiny, instantaneous dipole, which can also induce an opposing dipole in neighbouring molecules. This leads to weak intermolecular attractions which are known as *dispersive forces* or *London forces*, and are present in all molecules, ions and atoms – even those with no permanent dipole moment. Dispersive forces decrease rapidly with distance, and the attractions are in proportion to $1/r^6$, where r is the distance between attracting species.

1.4.2 Dielectric Constant

Dielectric constant (or *relative permittivity*), ε_r, is an indication of the polarity of a solvent, and is measured by applying an electric field across the solvent between

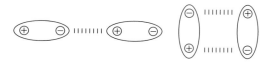

Figure 1.9 Possible alignments of two dipolar molecules that can lead to attraction and short-range structuring of liquids

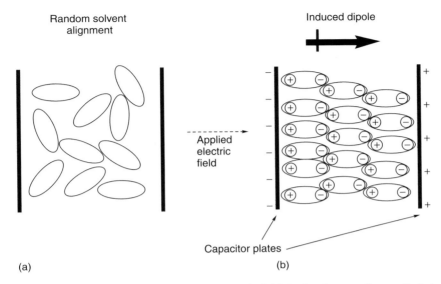

Figure 1.10 Polarization of molecules in an electric field. In the absence of an applied electrical field (a), molecules are aligned randomly, with no net dipole. When the field is applied (b), the solvent molecules are polarized and align themselves to reduce the strength of the field

two plates of a capacitor. The electric field will induce a dipole opposite to the applied field, even in solvents with no permanent dipole of their own, as shown in Figure 1.10. This phenomenon is often referred to as *polarizability*. Because this behaviour is similar to the orientation of a solvent around an electrolyte, ε_r is usually a good indicator of the ability of the solvent to dissolve ionic compounds. The dielectric constant is measured relative to the effect of the same applied field when applied to a vacuum (Equation 1.3), where ε_o is the permittivity of a vacuum.

$$\varepsilon_r = \frac{\varepsilon_{solvent}}{\varepsilon_o} \tag{1.3}$$

1.4.3   Electron Pair Donor and Acceptor Numbers

1.4.3.1   Donor Number, DN

The *donor number*, DN, of a solvent, proposed by Gutmann, is a measure of the Lewis basicity of the solvent, i.e. its ability to donate a pair of electrons [16]. The DN is determined by measuring the negative enthalpy for the reaction of equimolar quantities of the solvent with the standard Lewis acid, $SbCl_5$, at room temperature in 1,2-dichloroethane (Scheme 1.1), and reflects the ability of the solvent to solvate Lewis acids. $SbCl_5$ reacts with protic solvents such as alcohols

Polar and Nonpolar Solvents

The terms polar, apolar and dipolar are frequently used to describe solvents and other molecules, but there is a certain amount of confusion and inconsistency in their application. *Dipolar* is used to describe molecules with a permanent dipole moment. *Apolar* should only be used to refer to solvents with a spherical charge distribution. All other solvents should be considered polar. Strictly speaking, by this definition, compounds such as carbon tetrachloride and benzene which are not spherical and may be polarized in an electrical field (see section on dielectric constant), are polar, and this polarizability is important when explaining the properties of those solvents. However, they do not have a permanent net dipole moment and give low values on most scales of solvent polarity. They are widely, if erroneously, termed *nonpolar*, and, although misleading, this name is useful in distinguishing solvents of low polarity from those with permanent dipoles. Solvents that are able to donate an acidic hydrogen to form a H-bond are termed *protic*, and those that cannot are called *aprotic*.

Solvent Classes and Examples

Apolar	liquid xenon, neon, argon
Nonpolar/polarizable	carbon tetrachloride, benzene, cyclohexane
Dipolar aprotic	chloroform, dimethylformamide, dimethyl sulfoxide
Dipolar protic	ethanol, methanol, water, liquid ammonia

and water, and the DN for these must be estimated by indirect methods. As the DN scale uses non-SI units, values of DN are usually normalized using 1,2-dichloroethane ($DN = 0 \, kcal \, mol^{-1}$) and hexamethylphosphoramide (HMPA, $DN = 38.8 \, kcal \, mol^{-1}$) as reference solvents for the scale. This generates the parameter DN^N (which corresponds to DN/38.8), thereby giving a unitless scale on which most solvents fall between 0 and 1 [17][2]. Table 1.6 shows values of donor number and other properties for some representative solvents.

$$\text{Solvent:} + SbCl_5 \longrightarrow \text{solvent} \longrightarrow SbCl_5$$

Scheme 1.1

---

[2] In fact, tris(pyrrolidino)phosphane oxide has a donor number of 1.22.

Table 1.6 Dipole moments (μ), dielectric constants (ε_r), normalized donor numbers (DN^N) and acceptor numbers (AN) for some common solvents [1, 2]

Solvent	$\mu(10^{-30}\,C\,m)$	ε_r	DN^N	AN
Water	5.9	78.3	0.46	54.8
Methanol	5.7	32.7	0.77	41.5
Ethanol	5.8	24.6	0.82	37.9
Dimethyl sulfoxide	13.5	46.5	0.77	19.3
N,N-dimethylformamide	12.4	37.8	0.69	16.0
Acetonitrile	11.8	35.9	0.36	18.9
Acetone	9.0	20.6	0.44	12.5
Ethyl acetate	6.1	6.0	0.44	9.3
Tetrahydrofuran	5.8	7.6	0.52	8.0
Diethyl ether	3.8	4.2	0.49	3.9
Dichloromethane	5.2	8.9	0.03	20.4
Chloroform	3.8	4.8	0.10	23.1
Carbon tetrachloride	0.0	2.2	0	8.6
Benzene	0.0	2.3	<0.01	8.2
Toluene	1.0	2.4	<0.01	–
Cyclohexane	0.0	2.0	0	0
n-Hexane	0.0	1.9	0	0

1.4.3.2 Acceptor number, AN

The *acceptor number*, AN, of a solvent is a measure of the power of the solvent to accept a pair of electrons [18]. Experimental evaluation of AN involves observing the frequency changes induced by a solvent on the ^{31}P NMR spectrum when triethylphosphine oxide, $Et_3P{=}O$, is dissolved in the solvent. Donation of an electron pair from the oxygen atom of $Et_3P{=}O$, as shown in Scheme 1.2, reduces the electron density around the phosphorus, causing a deshielding effect which leads to an increase in chemical shift. Hexane (AN = 0) and $SbCl_5$ (AN = 100) were used as fixed points to define this scale.

1.4.4 Empirical Polarity Scales

Quantitative determination of solvent polarity is difficult, and quantitative methods rely on physical properties such as dielectric constant, dipole moment and refractive index. It is not possible to determine the solvent polarity by measuring an individual solvent property, due to the complexity of solute–solvent interactions, and for this reason empirical scales of solvent polarity based on chemical

$$Et_3P{=}O + Solvent \longrightarrow Et_3P{-}O{-}Solvent$$

Scheme 1.2

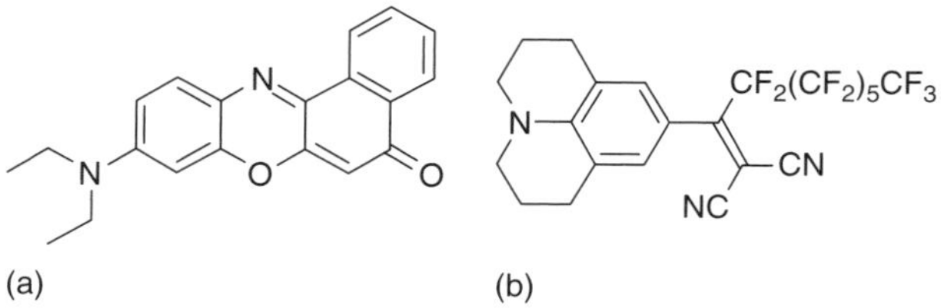

(a)   (b)

Figure 1.11   Solvatochromic dyes: Nile Red (a) and an α-perfluoroalky-β,β-dicyanovinyl compound (b)

properties are widely used instead. It is preferable that these methods are easy to conduct from an experimental point of view. The most common method is to dissolve a compound with some solvent sensitive spectroscopic characteristic, a phenomenon known as *solvatochroism*. The use of solvatochromic dyes, which undergo a change in their UV spectrum in different solvents, has become a very popular way of achieving this. Some of these empirical methods are discussed below, but it should be noted that many other indicators of solvent polarity have been proposed and used with success [1]. Nile Red has been used to measure the polarity of some ionic liquids (see Chapter 4), and an α-perfluoroalky-β,β-dicyanovinyl dye was successfully used to produce a spectral polarity index (P_s) for fluorinated solvents in which other, more widely known probe molecules were not sufficiently soluble to be used (see Chapter 3) [19]. These dyes are shown in Figure 1.11.

1.4.5 E_T^N and $E_T(30)$ Parameters

Dimroth and Reichardt devised a polarity scale based on the solvatochromic behaviour of the pyridinium-N-phenoxide betaine dye shown in Figure 1.12 [20]. This dye is the most solvatochromic compound reported to date, showing a range of transition energies for the $\pi \rightarrow \pi^*$ absorption band from $147\,\text{kJ}\,\text{mol}^{-1}$ ($810\,\text{nm}$) in diphenyl ether to $264\,\text{kJ}\,\text{mol}^{-1}$ ($453\,\text{nm}$) in water [1, 21]. This exceptional behaviour makes the dye a useful indicator of solvent polarity, with the measurement being made by human eye (the absorption range is almost entirely within the visible region), or, more quantitatively, by UV-vis spectroscopy. The dye is green in dichloromethane, purple in ethanol and red in methanol, as shown in Figure 1.13.

The original solvent polarity scale, known as the $E_T(30)$ scale, was defined simply as the energy of the longest wavelength adsorption band for the dye, measured in $\text{kcal}\,\text{mol}^{-1}$. This scale has now been revised and normalized because of the introduction of SI units, and E_T^N is defined in Equation 1.4.

$$E_T^N = \frac{E_T(\text{solvent}) - E_T(\text{TMS})}{E_T(\text{water}) - E_T(\text{TMS})} \tag{1.4}$$

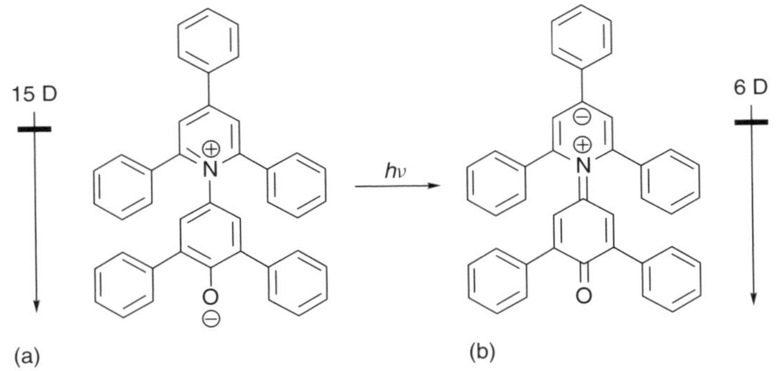

Figure 1.12 Reichardt's betaine dye in its zwitterionic ground state (a) and first excited state (b). The ground state has a larger dipole moment (15 D) than the excited state (6 D). Measurement of the energy of the transition between these two states ($\pi \rightarrow \pi^*$) is the basis for the E_T^N scale of solvent polarity

Figure 1.13 (Plate 1) Solvatochromic behaviour of Reichardt's dye, in (from left to right) dichloromethane, acetone, acetonitrile, ethanol and methanol

where E_T is the energy for the $\pi \rightarrow \pi^*$ transition of the dye in a given solvent. Tetramethylsilane (TMS), by definition, has a E_T^N value of zero. Some E_T^N values for common solvents are given in Table 1.7.

More polar solvents, such as water and acetonitrile, stabilize the charged zwitterionic ground state of the dye more than the less dipolar excited state, which leads to a larger energy change for the $\pi \rightarrow \pi^*$ transition than in less polar media, as shown in Figure 1.14. In this application the dye is acting as a *reporter* molecule, revealing information about its local environment via an easy-to-measure property.

Although Reichardt's dye itself is not soluble in solvents of very low polarity (diphenyl ether is the lower limit), a related dye functionalized with *t*-butyl groups which has increased solubility in hydrocarbons, shown in Figure 1.15, has been used to extend the scale. However, attempts to use a similar approach to gain

Table 1.7 $E_T(30)$ and E_T^N values for some common solvents

Solvent	$E_T(30)(\text{kcal mol}^{-1})$	E_T^N
Water	63.1	1.000 (defined)
Methanol	55.4	0.762
Ethanol	51.9	0.654
Acetonitrile	45.6	0.460
Dimethyl sulfoxide	45.1	0.444
N,N-dimethylformamide	43.8	0.404
Acetone	42.2	0.355
Dichloromethane	40.7	0.309
Chloroform	39.1	0.259
Ethyl acetate	38.1	0.228
Tetrahydrofuran	37.4	0.207
Diethyl ether	34.5	0.117
Benzene	34.3	0.111
Toluene	33.9	0.099
Carbon tetrachloride	32.4	0.052
n-Hexane[a]	31.0	0.009
Cyclohexane[a]	30.9	0.006
Tetramethylsilane[a]	30.7	0.000 (defined)

[a] Measured using the t-butyl-substituted dye shown in Figure 1.15.

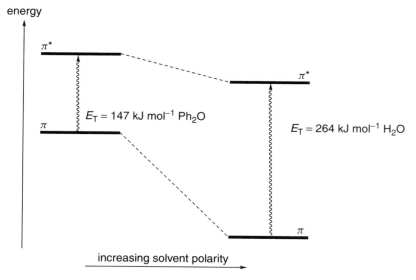

Figure 1.14 Increasing solvent polarity stabilizes the zwitterionic ground state (π) of Reichardt's dye relative to the first excited state (π^*), leading to an increase in the transition energy

$R_f = CF_3, C_6F_{13}$

(a)   (b)

Figure 1.15 *t*-Butyl- (a) and perfluoroalkyl- (b) substituted betaine dyes for polarity measurements in less polar or highly fluorous solvents

useful measurements of fluorinated solvents by attaching fluorinated chains to the dye have proved largely unsuccessful [22].

As the E_T^N value is related to the ability of a solvent to stabilize charge separation in the dye, a correlation between E_T^N and dielectric constant, ε_r might be expected. Figure 1.16 shows a plot of E_T^N against ε_r for the solvents shown in Table 1.7. Not all solvents show a good correlation, which highlights the difficulty of producing a single parameter capable of describing solvent polarity. Dipolar, aprotic solvents appear to give a lower E_T^N value than would be expected on the basis of their dielectric constants. It should be remembered that the dye might form specific Lewis acid/base interactions with solvent molecules, as well as being sensitive to bulk, nonspecific solvent 'polarity'. A more detailed discussion of the relationships between empirical solvent scales may be found elsewhere [2].

1.4.6 Kamlet–Taft Parameters

From a practical viewpoint, E_T^N values are quickly and easily obtained, giving a very useful and convenient scale. However, a general polarity scale based on a single probe molecule has its limitations because a single compound cannot experience the diversity of interactions that the whole range of solvents can offer. The Kamlet–Taft parameters α, β and π^* tackle this problem by using a series of seven dyes to produce a scale for specific and nonspecific polarity of liquids [23]. Whilst it undoubtedly gives a more detailed description of the solvents properties,

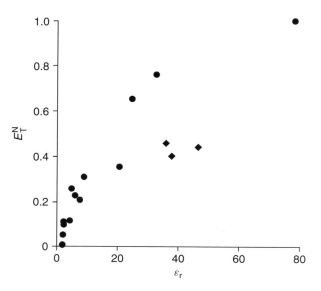

Figure 1.16 Relationship between E_T^N and ε_r for the solvents listed in Table 1.7. ◆, dimethyl sulfoxide, N,N-dimethylformamide and acetonitrile; ●, other solvents

Table 1.8 Kamlet–Taft α, β and π^* parameters for selected solvents [1, 2]

Solvent	α	β	π^*
Water	1.17	0.47	1.09
Methanol	0.98	0.66	0.60
Ethanol	0.86	0.75	0.54
Dimethyl sulfoxide	0.00	0.76	1.00 (defined)
N,N-dimethylformamide	0.00	0.69	0.88
Acetonitrile	0.19	0.40	0.75
Acetone	0.08	0.43	0.71
Ethyl acetate	0.00	0.45	0.55
Tetrahydrofuran	0.00	0.55	0.58
Diethyl ether	0.00	0.47	0.27
Dichloromethane	0.13	0.10	0.82
Chloroform	0.20	0.10	0.58
Carbon tetrachloride	0.00	0.10	0.28
Benzene	0.00	0.10	0.59
Toluene	0.00	0.11	0.54
n-Hexane	0.00	0.00	−0.04
Cyclohexane	0.00	0.00	0.00 (defined)
Perfluoromethylcyclohexane	0.00	−0.06	−0.40

this approach is more time consuming and requires extra measurements and calculations. The Kamlet–Taft equation is generally represented as follows:

$$X = X_o + a\alpha + b\beta + s\pi^* \tag{1.5}$$

where X_o, a b and s are solvent-independent constants for the solvatochromic indicator under study. X is the empirical measurement, i.e. the solvatochromic shift of the dye. α is the *hydrogen bond donor* (HBD) ability of the solvent and β is the *hydrogen bond acceptor* (HBA) ability. HBD and HBA solvents are discussed briefly below in Section 1.4.7. The parameter π^* is therefore a measure of the residual general polarity/polarizability of the solvent after H-bonding effects have been removed. Values for α, β and π^* have been obtained for a wide range of solvents which makes it an extremely useful scale for comparative purposes [1, 2]. Some Kamlet–Taft values for common solvents are given in Table 1.8.

1.4.7 Hydrogen Bond Donor (HBD) and Hydrogen Bond Acceptor (HBA) Solvents

Hydrogen bond donor solvents are simply those containing a hydrogen atom bound to an electronegative atom. These are often referred to as protic solvents, and the class includes water, carboxylic acids, alcohols and amines. For chemical reactions that involve the use of easily hydrolysed or solvolysed compounds, such as $AlCl_3$, it is important to avoid protic solvents. Hydrogen bond acceptors are solvents that have a lone pair available for donation, and include acetonitrile, pyridine and acetone. Kamlet–Taft α and β parameters are solvatochromic measurements of the HBD and HBA properties of solvents, i.e. acidity and basicity, respectively [24]. These measurements use the solvatochromic probe molecules N,N-diethyl-4-nitroaniline, which acts as a HBA, and 4-nitroaniline, which is a HBA *and* a HBD (Figure 1.17).

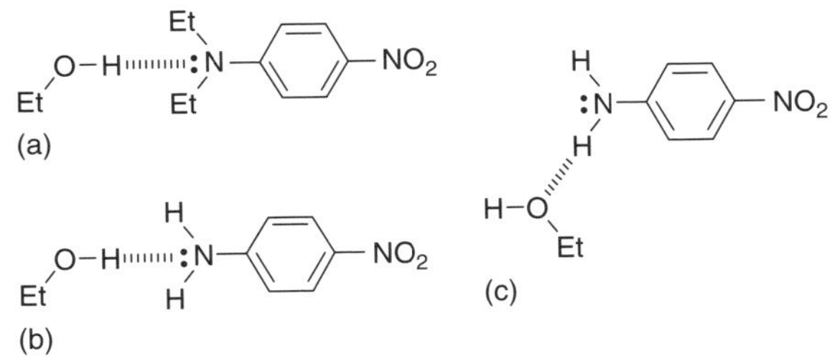

Figure 1.17 Hydrogen-bond formation between nitroanilines and ethanol. (a) N,N-diethyl-4-nitroaniline is a HBA only; (b) 4-nitroaniline is a HBA and (c) a HBD

1.5   THE EFFECT OF SOLVENT POLARITY ON CHEMICAL SYSTEMS

The main principles and concepts of the effect of solvent polarity on chemical reactions and equilibria are outlined in the following sections. However, this is a vast subject area beyond the scope of this work and the interested reader will find a detailed discussion elsewhere [1].

1.5.1   The Effect of Solvent Polarity on Chemical Reactions

Generally, the effect of changing the solvent in which a chemical reaction is conducted may be understood by considering the charge distribution of the reactants and transition states. In each case, a more polar solvent will lead to stabilization of species with dipoles or charge separation, relative to nonpolar species. For example, in a typical S_N1 nucleophilic substitution reaction, where an uncharged reactant forms a charged transition state, a more polar solvent will stabilize the transition state complex relative to the reactant, lowering the activation enthalpy, $\Delta H^{\ddagger}$, as shown in Figure 1.18. For example, the S_N1 solvolysis of t-butyl chloride (2-chloro-2-methylpropane), in which the rate limiting step is the heterolytic breaking of the C–Cl bond, is 3.4×10^5 times faster in water than in less polar ethanol [25].

By contrast, in an S_N2 reaction, a charged nucleophile reacts with an uncharged substrate to form a transition state in which the negative charge of the nucleophile

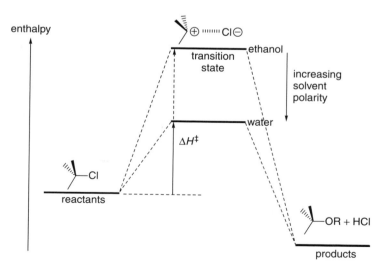

Figure 1.18   Solvent polarity effect in an S_N1 reaction. Increasing the polarity of the solvent stabilizes the charge separation formed in the transition state, lowering the activation energy and increasing the rate of reaction

is dispersed over the entire complex. The charge separation in the transition state is reduced compared with the reactants. In this situation, the use of a polar solvent will stabilize the starting material relative to the transition state, as shown in Figure 1.19. The reaction will thus proceed more slowly in solvents of higher polarity.

The direction and extent of the effect of solvent polarity on reaction rates of nucleophilic substitution reactions are summarized by the Hughes–Ingold rules, shown in Table 1.9 [26]. These rules do not account for the entropic effects or any specific solvent–solute interactions such as H-bonding, which may lead to extra stabilization of reactants or transition states [27].

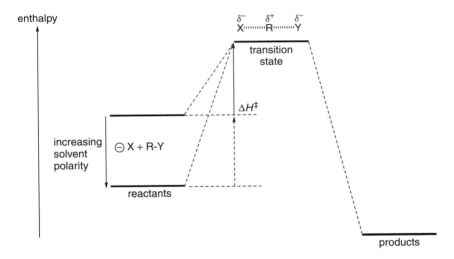

Figure 1.19 The effect of solvent polarity in an S_N2 reaction. Increasing the polarity of the solvent stabilizes reactants relative to the transition state, raising the activation energy and decreasing the rate of reaction

Table 1.9 Hughes–Ingold rules for solvent effects in nucleophilic substitution reactions [27, 28]

Reaction	Reactants	Transition state	Change in charge distribution	Effect of increasing solvent polarity	Size of effect
S_N2	$Y^- + R–X$	$^{\delta-}Y\ldots R\ldots X^{\delta-}$	Dispersed	Decrease	Small
S_N2	$Y + R–X$	$^{\delta+}Y\ldots R\ldots X^{\delta-}$	Increased	Increase	Large
S_N2	$Y^- + R–X^+$	$^{\delta-}Y\ldots R\ldots X^{\delta+}$	Decreased	Decrease	Large
S_N2	$Y + R–X^+$	$^{\delta+}Y\ldots R\ldots X^{\delta+}$	Dispersed	Decrease	Small
S_N1	$R–X$	$^{\delta+}R\ldots\ldots X^{\delta-}$	Increased	Increase	Large
S_N1	$R–X^+$	$^{\delta+}R\ldots\ldots X^{\delta+}$	Dispersed	Decrease	Small

Neutral reactants that form neutral products by passing through an uncharged transition state are expected to show little or no solvent polarity-dependent changes in reaction rate. This category of reaction includes pericyclic reactions and rearrangements, including the Diels–Alder reaction. However, dramatic changes in reaction rates for such reactions have been reported for such reactions when using either very polar *or* very nonpolar solvents. Although the solvent has no charge-stabilizing effect on the transition state in these cases, other solvent phenomena, including the *hydrophobic effect*, can lead to enhanced rates. These ideas are discussed further in Chapters 5 and 7.

1.5.2   The Effect of Solvent Polarity on Equilibria

Like reaction rates, the effect of solvent polarity on equilibria may be rationalized by consideration of the relative polarities of the species on each side of the equilibrium. A polar solvent will therefore favour polar species. A good example is the *keto–enol* tautomerization of ethyl acetoacetate, in which the 1,3-dicarbonyl, or *keto*, form is more polar than the *enol* form, which is stabilized by an intramolecular H-bond. The equilibrium is shown in Scheme 1.3. In cyclohexane, the *enol* form is slightly more abundant. Increasing the polarity of the solvent moves the equilibrium towards the *keto* form [28]. In this example, H-bonding solvents will compete with the intramolecular H-bond, destabilizing the *enol* form of the compound.

Solvent polarity effects are also seen in the formation of isomers of transition metal complexes. Reactions that give a mixture of *cis* and *trans* isomers can be tuned by careful choice of solvent to give one isomer in preference to the other. For example, with *cis* and *trans*-[Pt(H$_2$L-S)$_2$Cl$_2$] (where H$_2$L = *N*-benzoyl-*N'*-propylthiourea), shown in Scheme 1.4 [29], the *cis* isomer is favoured in solvents of high polarity whereas the *trans* isomer is dominant in solvents of low polarity. These observations are in accordance with other related observations [30], and

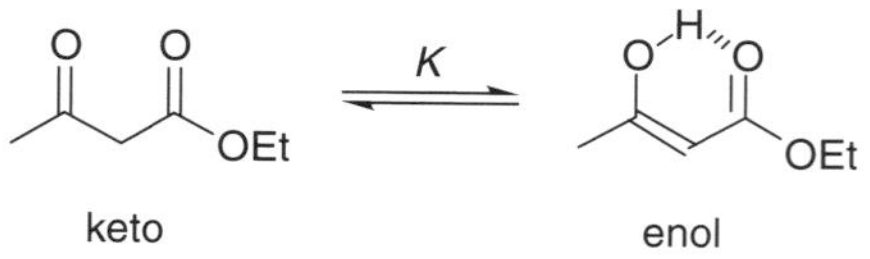

	Solvent	K
$K = \dfrac{[enol]}{[keto]}$	Cyclohexane	1.07
	Diethylether	0.54
	Tetrahydrofuran	0.33
	Ethanol	0.18
	Water	0.06

Scheme 1.3   The effect of solvent polarity on *keto–enol* tautomerization of ethyl acetoacetate

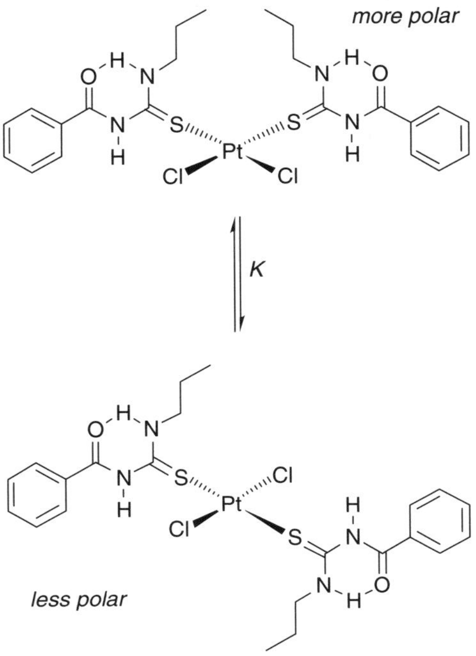

Scheme 1.4 The more polar *cis* form of the platinum complex is dominant in polar solvents

represent a case of 'like yielding like' since the resulting *cis* complex is more polar than the *trans* complex. From the data in Table 1.10, it is apparent that the position of the equilibrium is highly solvent dependent. The relative rates of isomerization are greatest in nonpolar solvents and as polarity is increased the rate of isomerization decreases. With complexes that exist in solution as an equilibrium *cis–trans* mixture, it has been possible to crystallize the *cis* isomer from a low polarity solvent and the *trans* isomer from a polar solvent [31].

Table 1.10 Dependence of equilibrium from Scheme 1.4 on solvent polarity [31]

Solvent	ε_r	E_T^N	Equilibrium constant K
Benzene	2.27	0.111	0.88
Chloroform	4.70	0.259	0.47
Tetrahydrofuran	7.39	0.207	0.39
Acetone	20.5	0.355	0.28
N,N-dimethylformamide	36.7	0.404	0.23
Nitromethane	38.6	0.481	0.16

1.6 WHAT IS REQUIRED FROM ALTERNATIVE SOLVENT STRATEGIES?

The problem with solvents is not so much their use, but the inefficiencies associated with their recovery and reuse. High volatility, whilst being an extremely useful property, leads to solvent losses to the environment. If a process consists of a reaction stage and a purification stage, solvents may be used and lost at each stage, as shown schematically in Figure 1.20a. Real chemical processes may include several separation steps, with further opportunities for solvent loss.

Alternative solvent strategies should allow efficient recovery and reuse of the solvent if environmental damage is to be avoided. Because there is a huge variety of solvents with a wide spectrum of physical properties, a range of alternatives to cover the full range of these different physical characteristics is required. It is unlikely that there is one, simple approach that will work in all applications, and what might be economically appropriate for a high value pharmaceutical might prove impractical for a bulk commodity.

Exceptions to this rule of high recovery may be made for CO_2 and water, which are non-toxic and may be returned to the environment providing that

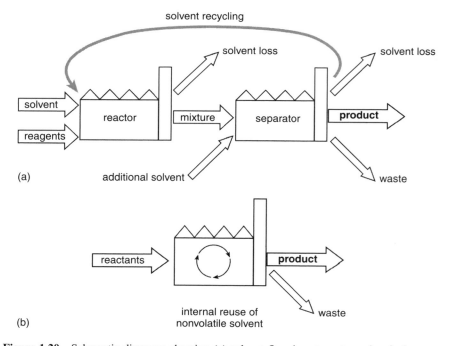

Figure 1.20   Schematic diagrams showing (a) solvent flow in a two-stage chemical process, where volatile organic solvents may be lost in both the reaction and separation steps; (b) improved efficiency in a process using a nonvolatile replacement solvent to reduce losses and improve separation

they are not contaminated by other, less benign chemicals. Although CO_2 is a greenhouse gas, and guidelines for the reduction of its emission have been set in law, its use as a solvent is not viewed as damaging. When used as a solvent, CO_2 is collected from air (or as a by-product of fermentation), converted to a liquid or supercritical fluid, and then returned to the atmosphere after use, giving a net change of zero. Of course, energy is required to compress the CO_2 and perform the phase change (which, ironically, may come from the burning of fossil fuels which *does* produce a net gain in CO_2), but this may be offset against the improvement in efficiency of any separation steps involved in the process. As CO_2 is a gas at ambient temperature and pressure, it can be removed simply by releasing the pressure from the reactor, to leave reaction products completely free of solvent and no energy is required to distil the solvent. The use of supercritical CO_2 and other supercritical fluids as reaction media is covered in Chapter 6.

If nonvolatile liquids are to be used to avoid the problems associated with volatile organic solvents, then it is very desirable that there is some convenient way of recovering the reaction products from the liquid. This approach is used in the biphasic systems described in Chapters 2–5. In the fluorous biphase (Chapter 3), reagents and catalysts are fine-tuned by adding perfluoroalkyl chains, known as 'ponytails', to ensure that only those chemicals will mix with the fluorous layer. Purification is simply a matter of separating the two phases. Transition metal catalysts with fluorous ligands will remain in the fluorous phase, and the whole catalyst–solvent mixture may be reused for another batch of reactions, as shown schematically in Figure 1.20b.

Ionic liquids may be used in a similar fashion, but in contrast to the extremely nonpolar fluorous solvents, ionic liquids are polar. They are completely non-volatile and so cannot be lost to the atmosphere. A range of ionic compounds that are liquid at room temperature and their use in synthetic chemistry are described in Chapter 4.

Water is abundant and nontoxic and is therefore a desirable solvent in environmental terms. However, cleaning up water that is contaminated with trace amounts of organic or metal catalyst impurities is very expensive. Water may be used as the sole solvent for some organic reaction chemistry, or in biphasic reactions alongside organic, fluorous, or ionic liquid solvents. Phase transfer catalysis (PTC) allows the use of inorganic reagents for reactions with organic substrates without the need for a polar, nonaqueous mutual solvent. The use of water as a solvent and PTC are covered in Chapter 5.

If these new solvent technologies are to be implemented for real industrial scale chemical synthesis, they must give some genuine advantage over existing methods, in terms of overall cost, safety, product quality, or some other improvement. PTC is already in use in manufacturing processes [32], because it allows the use of cheap, inorganic reagents and reduces or eliminates the need for additional organic solvent. Water is used as a solvent for industrial hydroformylation reactions [33]. Supercritical carbon dioxide (sc CO_2) is used for decaffeination of coffee beans and for the extraction of bioactive molecules from crops [34].

A large scale supercritical reactor for chemical synthesis has recently been built in the UK and it seems likely that more synthetic processes using CO_2 as a solvent will appear in the near future [35].

At the time of writing, neither ionic liquids nor fluorous solvents have been used as solvents for commercial processes, although BASF use an ionic liquid as a proton scavenger in phosphine manufacture [36]. These are relatively new technologies and time will tell whether or not they will fulfil current high expectations.

Whilst the first part of this book deals mainly with concepts and theories of alternatives to the use of organic solvents as reaction media, the later Chapters (Chapters 7–11) look at some applications of these methods and compare their strengths and limitations as illustrated by these case studies.

REFERENCES

1. Reichardt C. *Solvents and Solvent Effects in Organic Chemistry, 2nd Edition*, Wiley-VCH, Weinheim, 1988.
2. Marcus Y. *Chem. Soc. Rev.* 1993, 409.
3. (a) Bell S. and McGillavray D. *Environmental Law*, 5th Edition, Blackstone Press, London, 2000; (b) McEldowney J. F. and McEldowney S. *Environmental Law and Regulation*, Blackstone Press, London, 2001; (c) Hester R. E. and Harrison R. M. (eds), *Global Environmental Change*, Royal Society of Chemistry, Cambridge, 2002; (d) Clarke A. G. In *Understanding Our Environment: An Introduction to Environmental Chemistry and Pollution*, Harrison R. M. (ed.), Royal Society of Chemistry, Cambridge, 1992, p. 5.
4. (a) EC directive 96/61 concerning Integrated Pollution Prevention and Control (IPPC), 1996; (b) Pollution Prevention and Control Act (UK), 1999; (c) The Pollution Prevention and Control (Scotland) Regulations, SSI 323, 2000; (d) The Pollution Prevention and Control (England and Wales) Regulations, SI 1973, 2000; (e) The Pollution Prevention and Control (England and Wales) Regulations Amendment, SI 275, 2002.
5. (a) US Environmental Protection Agency, Environmental Finance Advisory Board/ Environmental Finance Centre, Guidebook, April 1999; (b) European Environment Agency, *Late Lessons from Early Warnings: The Precautionary Principle 1896–2000*, Copenhagen, 2001.
6. Riley C. *Purchasing Magazine* 7 October 1999.
7. Landau P. *Chemical Market Reporter* 7 June 1999.
8. Cage S. *Chemical Week* 19 June 2002.
9. Matlack A. S. *Introduction to Green Chemistry*, Marcel Dekker, New York, 2001, p. 201.
10. European Chlorinated Solvent Association, Sales data, June 2002.
11. (a) Baghurst D. R. and Mingos D. M. P. *J. Chem. Soc., Chem. Commun.* 1992, 674; (b) Gabriel C., Gabriel S., Grant E. H., Halstead B. S. J. and Mingos D. M. P. *Chem. Soc. Rev.* 1998, **27**, 213.
12. (a) Correa W. H., Edwards J. K., Macluskey A., McKinnon I., and Scott J. L. *Green Chem.*, 2003, **5**, 30; (b) Subhas Bose D., Fatima L. and Babu Mereyala H. *J. Org. Chem.*, 2003, **68**, 587; (c) Tanaka K. and Toda F. *Chem. Rev.* 2000, **100**, 1025.
13. Hussey C. L., In *Advances in Molten Salts*, Mamantov G. and Mamantov C. (eds), Elsevier, New York, 1983, Vol. 5, p. 185.
14. Tanaka K. and Toda F. *Chem. Rev.* 2000, **100**, 1025.

15. Reichardt C. *Nachr. Chem. Tech. Lab.* 1997, **45**, 759.
16. Fawcett W. R. *J. Phys. Chem.* 1993, **97**, 9540.
17. Reichardt C. In *Organic Liquids: Structure, Dynamics, and Chemical Properties*, Buckingham A. D., Lippert E. and Bratos S. (eds), Wiley, Chichester, 1978, pp. 269–281.
18. Mayer U., Gutmann V. and Gerger W. *Monatsh. Chem.* 1975, **106**, 1235.
19. (a) Carmichael A. J. and Seddon K. R. *J. Phys. Org. Chem.* 2000, **13**, 591; (b) Freed B. K., Biesecker J. and Middleton W. J. *J. Fluorine Chem.* 1990, **48**, 63.
20. Reichardt C. *Chem. Rev.* 1994, **94**, 2319.
21. Dimroth K., Reichardt C., T. Siepmann and Bohlmann F. *Liebigs Ann. Chem.* 1963, **661**, 1.
22. Reichardt C., Eschner M. and Schafer G. *J. Phys. Org. Chem.*, 2001, **14**, 737.
23. Kamlet M. J., Abboud J.-L. M., Abraham M. H. and Taft R. W. *J. Org. Chem.* 1983, **48**, 2877.
24. (a) Johnson B. P., Gabrielsen B., Matulenko M., Dorsey J. G. and Reichardt C. *Anal. Lett.* 1986, **19**, 939; (b) Reichardt C. and Schafer G. *Liebigs Ann. Chem.*, 1995, 1579.
25. Winstein S. and Fainberg A. F. *J. Am. Chem. Soc.* 1957, **79**, 5937.
26. (a) Hughes E. D. and Ingold C. K. *Trans. Faraday Soc.*, 1941, **37**, 603; (b) Ingold C. K. *Structure and Mechanism in Organic Chemistry*, Bell, London, 1953.
27. Burton C. A. *Nucleophilic Substitution at a Saturated Carbon Atom*, Elsevier, London, 1963.
28. Moriyasu M., Kato A. and Hashimoto Y. *J. Chem. Soc., Perkin Trans.* **2**, 1986, 515.
29. Koch K. R., Wang Y. and Coetzee A. *J. Chem. Soc., Dalton Trans.* 1999, 1013.
30. (a) Jenkins J. M. and Shaw B. L. *J. Chem. Soc. A* 1966, 770; (b) Redfield D. A. and Nelson J. H. *Inorg. Chem.*, 1973, **12**, 15.
31. Real J., Prat E., Polo A., Alvarez-Larena A. and Piniella J. F. *Inorg. Chem. Commun.* 2000, **3**, 221.
32. (a) Boswell C. *Custom Manufacturing* 4 November 2002; (b) Starks C. M. Liotta C. L. and Halpern M. *Phase Transfer Catalysis: Fundamentals, Applications and Industrial Perspectives*, Chapman & Hall, London, 1994.
33. (a) Kuntz E. G. French Patent no. 2314910, 1975 (to Rhone-Poulenc Recherche); (b) Cornils B. and Weibus E. *CHEMTECH* 1995, **25**, 33.
34. (a) Demirbas A. *Energy Conserv. & Manage.*, 2001, **42**, 279; (b) Dean J. R. and Khundker S. *J. Pharm. Biomed. Anal.* 1997, **15**, 875.
35. (a) Institute of Applied Catalysis. *iAc News* 2002, **9**, 1; (b) Press release PA66/02, University of Nottingham, 11 July 2002.
36. (a) Tzschucke C. C., Markert C., Bannwarth W., Roller S., Hebel A. and Haag R. *Angew. Chem., Int. Ed. Eng.* 2002, **41**, 3964; (b) Seddon K. *Green Chem.*, 2002, **4**, G25; (c) Guterman L. *The Economist* 19 June 1999, 123; (d) Freemantle M. *Chem. Eng. News*, 31 March 2003, 9.

2 Multiphasic Solvent Systems

One of the greatest challenges to chemists today is to replace existing technologies with cleaner processes and also to develop new products that are kinder to the environment. This requires a new approach, which sets out to reduce the materials and energy used in manufacturing process, to minimize, or ideally eliminate the dispersion of chemicals in the environment and maximize the use of resources, and to extend the durability and recyclability of products. There are numerous ways by which these goals can be achieved and one that is currently under intensive scrutiny is the development and implementation of alternative solvents to volatile organic compounds. The problems associated with conventional organic solvents were outlined in Chapter 1. Multiphasic synthetic methods have emerged as one of the most effective methods meeting these challenges as volatile organic solvent use is reduced or even completely eliminated from the process. In this chapter, the various approaches to biphasic, triphasic and multiphasic synthesis are described with attention paid to catalysed reactions. Following this discussion, the different solvent combinations that may be used are described. The advantages and disadvantages of multiphasic processes are assessed together with a discussion of partition coefficients and the methods to make a catalyst preferentially soluble in a particular solvent. Finally, following a brief revision of basic reaction kinetics, the kinetics of biphasic reactions are explored in relation to where the reaction in a biphasic system actually takes place.

2.1 AN INTRODUCTION TO MULTIPHASIC CHEMISTRY

There are numerous types of multiphasic chemical processes. The most common are biphasic although triphasic, tetraphasic and even higher number of phases can also be used to conduct chemical synthesis. All the multiphasic methods aim to overcome the major problem of homogeneous catalysis, which is catalyst recovery and product separation. The simplest systems are biphasic ones that involve immobilizing a catalyst in one solvent, which is immiscible with a second solvent in which the substrates/products are dissolved. If a gas is required as a substrate then the system could be regarded as triphasic (i.e. liquid–liquid–gas), although for the purposes of this book (and as is most commonly defined elsewhere) such as system will be referred to as biphasic. In other words, only the number of different liquid solvent phases will be used to define the phasicity of a system.

Chemistry in Alternative Reaction Media   D. Adams, P. Dyson and S. Tavener
© 2004 John Wiley & Sons, Ltd   ISBNs: 0-471-49848-3 (Cloth); 0-471-49849-1 (Paper)

Should a third solvent be added which is immiscible with the other two, then a triphasic system results. The purpose of a third solvent may be, for example, as a built in cleaning step (see below).

In general, there are four main types of biphasic processes, and they are generally used for catalysed reactions rather than stoichiometric ones. These four main processes are summarized below.

2.1.1 The Traditional Biphasic Approach

The most widely encountered biphasic method commences with two immiscible phases, one containing the catalyst, the other the substrate or substrates, and was first recognized by Manassen in 1973 [1]. Liquid phases may be immiscible if their polarities are sufficiently different, as explained in Chapter 1. The two phases are vigorously mixed allowing reaction between the catalyst and substrates to take place. When the reaction is complete, the mixing is stopped and the two phases separate. A schematic representation of such a process is illustrated in Figure 2.1. In the ideal system, the catalyst is retained in one phase ready for reuse and the product is contained in the other phase and can be removed without being contaminated by the catalyst. In certain cases, neat substrates may be used as one phase, without additional solvents.

There are several problems with this type of system. The first is that vigorous mixing is usually required in order to get high reaction rates, by increasing both the amount of substrate dissolved in the catalyst phase and the area of the interface between the two solvents. It should be noted that rapid mixing does not result in the formation of a homogeneous phase, but rather an emulsion in which the surface area between the two phases is maximized.

2.1.2 Temperature Dependent Solvent Systems

Certain solvents are essentially immiscible at low temperature, but on heating, their relative solubility increases until they form a single phase. Such a biphasic system allows reactions to be conducted in single phase under homogeneous

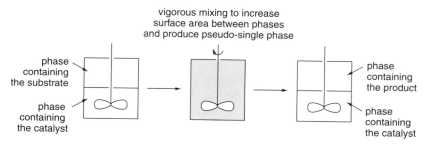

Figure 2.1 The classical biphasic process between two immiscible solvents

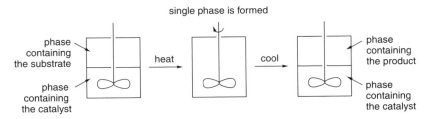

Figure 2.2   The temperature dependent biphasic process in which two phases are present at low temperature. These form a single phase on heating

conditions while maintaining all the advantages of biphasic extraction. The technique was first recognized in fluorous–organic biphasic catalysis [2], and has more recently been applied to ionic liquid–aqueous reactions. The approach is illustrated in Figure 2.2.

2.1.3   Single- to Two-Phase Systems

In some instances, the catalyst and substrates are soluble in the same solvent (especially when the substrates are gases), but the products are immiscible with the solvent and form a second phase as shown in Figure 2.3. If a reaction goes to completion, then it is not even necessary for the substrates to dissolve completely in the catalyst phase.

This type of process represents the ideal biphasic method as long as the product can be extracted without contamination from the catalyst and catalyst immobilization solvent. This technique is employed commercially for the production of butyraldehyde from propene, carbon monoxide and hydrogen which is described in detail in Chapter 11 [3].

2.1.4   Multiphasic Systems

In principle, there is no limit to the number of phases that can be used to conduct a chemical process. Triphasic processes have been shown to have some

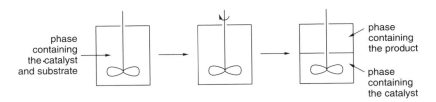

Figure 2.3   A single- to two phase system in which the substrates are initially dissolved in the same phase as catalyst and the product forms a new phase

advantages over biphasic processes, but there do not appear to be any multiphasic processes used in synthesis that go beyond three different liquid phases. A typical use for a third phase is as a type of built-in wash, removing unwanted side-products from the catalyst immobilization solvent that are also insoluble in the product phase.

An alternative triphase system has been developed in which a conventional organic solvent containing one reagent with the same organic solvent containing another reagent is separated by a different solvent phase of higher density [4]. The organic phase at the bottom diffuses through the higher density central phase into the organic phase at the top. The reagent also slowly diffuses and reacts in a controlled manner with the reagent in the upper phase. Only two phases remain once the reaction is complete. The technique has been termed a 'phase vanishing method' with the central phase acting as a type of liquid membrane.

2.2 SOLVENT COMBINATIONS

There are various combinations of solvents that lead to the different biphasic and triphasic systems described in Section 2.1 that have found applications in synthesis and catalysis. For practical purposes, it is essential that the catalyst and product

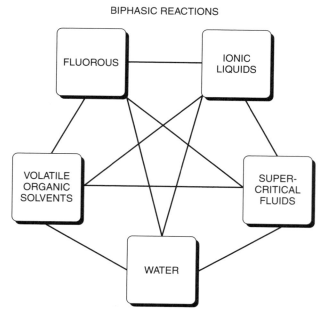

Figure 2.4 Fluid phases for biphasic chemistry: ——— indicates that a biphasic system may be formed

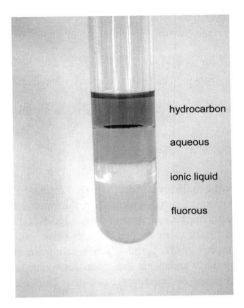

Figure 2.5 (Plate 2) Four liquid phases in a single test tube (note that the coloured phases have been dyed for clarity). (Photograph by Dave Adams; thanks to Glen Capper for supplying the ionic liquid)

phases effectively separate after reaction. The most important solvent combinations used in multiphasic chemistry are summarized in Figures 2.4 and 2.5. The key solvent combinations generally involve a conventional solvent combined with an alternative solvent, or two alternative solvents. The physical and chemical properties of these alternative solvents are described in greater depth in Chapters 3–6.

2.2.1 Water

Aqueous–organic biphasic catalysis is the most extensively studied biphasic method. Examples of reactions catalysed using the aqueous–organic approach include hydrogenations [5], hydroformylations [6], oxidations [7], C–C coupling [8], olefin metathesis [9] and polymerizations [10]. Although out of fashion for many years, stoichiometric organic reactions are also becoming increasingly common in water. The poor solubility of organic compounds in water, while being an advantage for product separation, can often cause considerable problems when it comes to bringing about the reaction. In general, as the temperature of the aqueous–organic biphase system is increased the solubility of organic substrates increases, but other protocols have also been developed to overcome solubility problems including:

1   Addition of co-solvents in which all the species are partially soluble.
2   Increasing the interface by rapid stirring, use of ultrasound, microwave dielectric heating, etc.
3   Use of detergents and surfactants.
4   Addition of phase transfer reagents, i.e. phase transfer catalysis.

Solubility problems in water and solutions to them are described in further detail in Chapter 5.

2.2.2   Fluorous Solvents

The miscibility of perfluoroalkanes and other perfluorinated solvents with hydrocarbon solvents is low, and this is exploited in fluorous–organic biphasic catalysis [11]. In some cases, apolar reactants may be dissolved in the fluorous phase and, on conversion to products of higher polarity, a second immiscible phase is formed. Notable examples of catalysed reactions that are effectively carried out using the fluorous biphase approach are hydroformylation and oxidation [12], although many other types of reactions have also been demonstrated. In general, these reactions are carried out using the second approach described in Section 2.1, i.e. where two phases are present at lower temperatures, but with the miscibility of the two solvents increasing with temperature until a point is reached where the two phases form a single homogeneous phase. The mixing problems sometimes encountered with traditional biphasic methods are thus removed, although other problems may be introduced, such as cross-contamination of the two phases during product extraction. However, a large number of different perfluorinated solvents are available, and careful solvent selection minimizes such problems.

2.2.3   Ionic Liquids

Many ionic liquids are immiscible with organic solvents and can be used in ionic liquid–organic catalysis. Since there is essentially no limit to the number of ionic liquids that can be made, it is possible to design ionic liquids with specific properties for a particular process [13]. Ionic liquid–organic biphasic processes have been assessed for the same range of reactions conducted using aqueous-organic solvents. Some ionic liquids are also immiscible, or partially immiscible, with water. Ionic liquid–aqueous processes based on the temperature dependent two-phase to one-phase biphasic approach have been demonstrated in hydrodimerization [14] and hydrogenation [15] reactions.

2.2.4   Supercritical Fluids and Other Solvent Combinations

In principle, many other solvent combinations could be used in biphasic chemistry although the main driving force in this area is to provide environmental benefits.

Supercritical fluids (e.g. supercritical carbon dioxide, scCO$_2$) are regarded as benign alternatives to organic solvents and there are many examples of their use in chemical synthesis, but usually under homogeneous conditions without the need for other solvents. However, scCO$_2$ has been combined with ionic liquids for the hydroformylation of 1-octene [16]. Since ionic liquids have no vapour pressure and are essentially insoluble in scCO$_2$, the product can be extracted from the reaction using CO$_2$ virtually uncontaminated by the rhodium catalyst. This process is not a true biphasic process, as the reaction is carried out in the ionic liquid and the supercritical phase is only added once reaction is complete.

A number of triphasic processes have been reported although the area remains underdeveloped, despite having considerable potential. For example, the epoxidation of *trans*-stilbene has been carried out using a fluorous–organic–aqueous system [17]. The catalyst is immobilized in the fluorous phase, the organic product is recovered from the organic phase and waste salt by-products are extracted into the aqueous phase. A triphasic system composed of ionic liquid, water and organic solvent has been used for Heck coupling reactions [18]. The catalyst is immobilized in the ionic liquid, the product is extracted into the organic phase and the water absorbs the inorganic salt by-products.

Although beyond the scope of this book, a vast amount of work has been directed to supporting homogeneous catalysts on solid supports including silica, alumina and zeolites, and functionalized dendrimers and polymers [19]. These give rise to so-called 'solid–liquid' biphasic catalysis and in cases where the substrate and product are both liquids or gases then co-solvents are not always required. In many ways solvent-free synthesis represents the ideal method but currently solvent-free methods can only be applied to a limited number of reactions [20].

2.3 BENEFITS AND PROBLEMS ASSOCIATED WITH MULTIPHASIC SYSTEMS

The most obvious benefit to using multiphasic reaction methods is that conventional organic solvent usage is reduced or even completely removed. Another very important aspect of biphasic catalysis can only be appreciated when the relative merits of homogeneous and heterogeneous catalytic methods are compared. Table 2.1 lists the main advantages and disadvantages of these two methods. It is worth noting that the majority of catalysed reactions by industry (which represents about 90 % of all reactions) use heterogeneous catalysts. However, homogeneous catalysts are becoming increasingly important, especially in enantioselective reactions.

From the data in Table 2.1, it is clear that homogeneous catalysts are superior to heterogeneous catalysts in almost every way except stability, separation and reuse (which are all linked). The importance of these properties explains the success of heterogeneous catalysts. Biphasic catalysis represents an intermediate process sandwiched somewhere between homogeneous and heterogeneous

Table 2.1 A comparison of the general properties of homogeneous and heterogeneous catalysis

Property	Homogeneous catalysis	Heterogeneous catalysis
Activity	High	Variable
Selectivity	High	Variable
Reaction conditions	Mild	Generally harsh
Stability/sensitivity	Low	High
Mixing	Easy	May be difficult
Catalyst separation	Difficult	Easy
Catalyst recycling	Difficult	Easy
Mechanistic understanding	Can be ascertained	Difficult to obtain
Catalyst tuning	By ligand modification	Difficult

catalysis. The catalyst is immobilized in a liquid and this allows, at least in principle, simple product extraction into a second phase, and also immediate reuse of the catalyst phase. In addition the lifetime of the catalyst may also improve. For example, product extraction from a homogeneously catalysed reaction often involves distillation of the product from the reaction mixture. This is not only expensive in terms of energy expenditure, but the temperatures involved may decompose the catalyst. Clearly, such separation procedures are not required for a biphasic process. In some cases, particularly with certain ionic liquids, the catalyst immobilization solvent also serves to protect the catalyst resulting in increased operation times and turnover numbers.

Despite the important advantages associated with biphasic processes, they are still not widely used on an industrial scale. The reason for the slow uptake of the technique stems from the fact that the immiscibility of two solvents is seldom perfect and some cross-contamination invariably takes place. It is therefore better to regard even biphasic systems as partially miscible liquids and the extent of miscibility depends strongly on temperature.

2.3.1 Partially Miscible Liquids

In the simplest binary system comprising two liquids (A and B), adding a small amount of either liquid to the other creates a single phase, as the one liquid dissolves completely in the other. As more of the second liquid is added, in this case B, the first liquid A becomes saturated with B and no more will dissolve. At this point, the system will consist of two phases in equilibrium with each other, one of liquid A saturated with B and the other of liquid B saturated with A. If B is continually added to A, there will come a point at which A becomes the minor component in the system and, ultimately, will dissolve completely in liquid B: a single phase will be formed once more. The relative proportions of each liquid that are required to form single or biphasic systems depends both

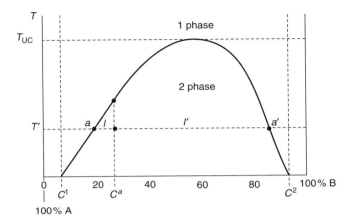

Figure 2.6 A temperature–composition diagram showing the relationship of temperature and solvent miscibility for two partially miscible liquids

on the physical properties of the two liquids, and on the temperature. The relationship between temperature, concentration and miscibility is described by the temperature–composition diagram shown in Figure 2.6.

The temperature–composition diagram can be used to calculate the composition of the two-phase system according to the amount of each solvent present. For example, at temperature T', the composition of the most abundant phase, which consists of liquid A saturated with liquid B, is represented by the point a and the composition of the minor phase, consisting of liquid B saturated with liquid A, is represented by point a'. The horizontal line connecting these two points is known as a tie line as it links two phases that are in equilibrium with each other. From this line the relative amounts of the two phases at equilibrium can be calculated, using the lever rule, under the conditions described by the diagram. The lever rule gets its name from a similar rule that is used to relate two masses on a lever with their distances from a pivot, i.e.:

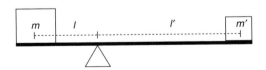

The lever rule states in order to balance the lever (i.e. create an equilibrium) then $ml = m'l'$ where m is the mass of the object and l is the length from the pivot.

The relative amounts of each phase in a biphasic system must also balance. So, to calculate their abundance a 'lever' is drawn along the line of the temperature, for example T' in Figure 2.6. The 'pivot' is the relative concentration of the two phases, so C^a, and the distances (l and l') are measured from the pivot to where the 'lever' crosses with the curve (a and a'). These values are related by Equation 2.1:

$$\frac{l'}{l} = \frac{\text{amount of phase of composition } a'}{\text{amount of phase of composition } a} \qquad (2.1)$$

If more of one of the liquids is added, the effect, according to the lever rule, is to shift the point of the pivot until balance is regained. Thus, at the given temperature the composition of the phases remains the same, i.e. each saturated with the other liquid, but the relative amount alters: if more of liquid B is added, then in the lever diagram the pivot point will shift to the right and thus more of this phase will form at the expense of the other phase (which is mainly liquid A). This rule applies to all partially miscible liquids.

The temperature also affects the composition of the two phases at equilibrium, but the effect is not equivalent in all systems. In the example shown in Figure 2.6, raising the temperature increases the solubility of the two phases and this is what is usually observed. The diagram shows that by heating the system, more of A dissolves in B and vice versa. However, other solvent pairs become less miscible with raised temperature, for example, water and ethylamine. In the case of these liquid pairs, the temperature–composition diagram is essentially reversed, as shown in Figure 2.7.

In both Figures 2.6 and 2.7 there is a temperature range in which phase separation does not occur. In Figure 2.6, this temperature (T_{UC}) is the maximum temperature at which two phases form, also known as the upper critical temperature. This maximum point occurs because the greater thermal motion in the system results in complete miscibility of the phases. In Figure 2.7, this temperature (T_{LC}) is the minimum temperature at which phase separation is observed, or

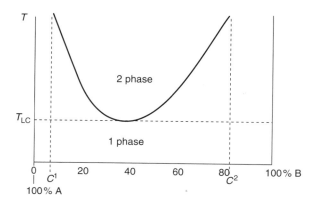

Figure 2.7 The temperature–composition diagram for a system with a lower critical temperature, such as water and ethylamine

the lower critical temperature. In the case of water and ethylamine, the minimum point occurs because below it the two components form a weak complex and therefore a single phase. Above this temperature, the thermal motion breaks up the complex and the two components are less miscible. There are also systems that have both upper and lower critical temperatures, where weak complexes form at lower temperatures and the thermal motion at higher temperatures causes complete miscibility. The temperature–composition diagrams for these systems have a circle area describing the two-phase system, with the single phase filling the outside of the circle in the diagram.

The consequence of incomplete phase separation in a biphasic catalysed reaction results in contamination of the product phase by some of the catalyst immobilization solvent, as well as the catalyst. In the worst possible case, a distillation process is still required to purify the product. In addition, with some of the catalyst lost from the immobilization phase (the catalyst is often expensive and toxic) the system is less active when a second batch of the substrate is introduced. The best way to minimize (or ideally eliminate) catalyst loss is to design a catalyst that is considerably more soluble in the immobilization phase compared to the product phase. This is usually done by attaching groups to the catalyst that provide the desired solubility properties for the immobilization solvent and many examples of these modified ligands are given in the following chapters.

2.4 KINETICS OF HOMOGENEOUS REACTIONS

Before discussing the kinetics of reactions in biphasic systems, the basics of kinetics in homogeneous reactions will be briefly revised. In all systems, the rate of a reaction corresponds to the amount of reactant that will be converted to product over a given time. The rate usually refers to the overall or net rate of the reaction, which is a result of the contributions of the forward and reverse reaction considered together. For example, consider the isomerization of *n*-butane to *iso*-butane shown in Scheme 2.1.

This reaction can be represented in a general way according to Equation 2.2.

$$a\text{A} \rightleftharpoons x\text{X} \qquad (2.2)$$

At the start of a reaction, when there is no or very little product present, the rate of the forward reaction is much greater than the reverse, such that the reverse reaction can be considered insignificant. These conditions are referred to as *pseudo*-first order kinetics. So, in the example, there is little or no *iso*-butane

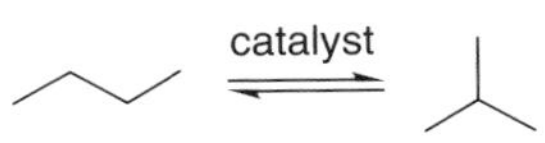

Scheme 2.1

at the start of the reaction. The rate of the reaction is therefore determined by the rate for isomerization of *n*-butane to *iso*-butane, and can be measured by the rate of change in the concentration of *iso*-butane with time, according to Equation 2.3.

$$\text{rate} = -\frac{1}{a}\frac{d[A]}{dt} \qquad (2.3)$$

where, for Scheme 2.1, A = *n*-butane. The square brackets denote the concentration of the component and *t* is time.

Essentially, what these terms mean is that for every small increase in time (dt), there is a change in the concentration of the reactant A (d[A]). For the reactant A, the concentration decreases as it is consumed by the reaction, thus the term is negative so that the rate of change calculated is positive. As the concentration of the product increases, the rate of change in concentration will be positive. In addition, each term is corrected for the stoichiometry of the reagent, by dividing the rate of change in concentration by the number of moles of each component used. With these equations, the value determined for the rate by monitoring the change in concentration of any of the components in the reaction (A or X in Equation 2.2) will be the same.

At the start of the reaction, the overall rate of change concentration of the reactant is linear with time. As the reaction proceeds and the product accumulates the reverse reaction becomes significant, such that the measured change in reagent concentration also decreases and the rate of the reaction is said to decrease. This decrease in rate is exponential, with the system eventually reaching equilibrium, where the amount of reactant converted to product equals the amount of product converted to reactant in a given time.

2.4.1   Rate is Independent of Stoichiometry

In a typical reaction, for example the isomerization of *n*-butane to *iso*-butane, the rate of the reaction can be described either by the disappearance of a reactant or the appearance of a product with time (Equation 2.3). The rate can also be determined by measuring the increase in concentration of the products, for example:

$$\text{rate} = \frac{1}{x}\frac{d[X]}{dt} \qquad (2.4)$$

Note that Equations 2.3 and 2.4 give the same value for rate, as the actual rate of change of each compound is corrected for stoichiometry. This becomes clear by considering the reaction shown in Equation 2.5, where two moles of methanol react with one mole of carbon dioxide and half a mole of oxygen to produce one mole each of carbonic acid dimethyl ester and water.

$$2CH_3OH + CO + 1/2O_2 \xrightarrow{\text{catalyst}} CH_3OCO_2CH_3 + H_2O \qquad (2.5)$$

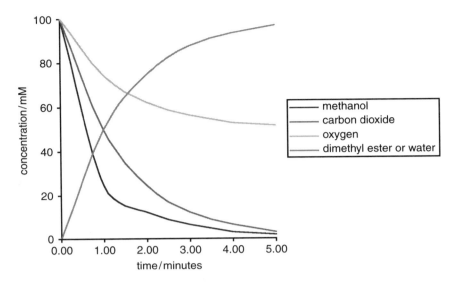

Figure 2.8 (Plate 3)   Change in concentration of reactant and products for the reaction shown in Equation 2.5

If the concentration of methanol is monitored with time, it will decrease twice as fast as carbon monoxide and four times as fast as oxygen as shown in Figure 2.8. However, by adjusting the change in concentration with time for the stoichiometry of the reaction, the determined rate of the reaction will be identical, no matter which of the reagents is monitored during the course of the reaction. Thus, the determined rate of reaction is independent of stoichiometry.

2.4.2   Rate is Determined by the Probability of Reactants Meeting

The rate of the reaction is related to probability of the reactants meeting in order to react. Therefore, the concentration of the reactants has an effect, because the probability of the reactants meeting is higher in a concentrated solution than in a dilute solution. Similarly, physical parameters such as agitation and temperature, that increase the rate of diffusion and molecular motion and therefore increase the probability of collisions, will also increase the rate of reaction.

2.4.3   Rate is Measured by the Concentration of the Reagents

At the start of a reaction, the reactants are present in large excess (*pseudo*-first order), and this simplifies the mathematical description of rate to that of a first order reaction, where rate is proportional to the concentration of a reactant or product. Thus, the two terms can be related by introducing a constant, k, termed the rate constant:

$$\frac{1}{a}\frac{d[A]}{dt} = k[A] \tag{2.6}$$

In a first order reaction, the change in concentration of the reactants or products with time is exponential (see Figure 2.8), and this can be illustrated mathematically by integrating and rearranging Equation 2.6 to give:

$$[A] = [A_o]e^{-kt} \tag{2.7}$$

Equation 2.7 shows that the concentration of reactant A at the start of the reaction will decrease exponentially with time as the reaction progresses. In relation to the concentration of the components of the reaction, this means that when there is a large excess of reactant, the rate of conversion to product will be much greater than when the concentration of the reactants is very low. Therefore, the more reactant present, the faster the reaction will proceed. However, note that the reaction described in Equation 2.2 proceeds in both directions, so that when the product begins to accumulate the reverse reaction will occur and equilibrium will be reached. Because of the complications in the mathematical description of the forward and reverse reactions occurring together, it is common to measure the initial rate of reaction, where there is little or no product and therefore the rate of the reverse reaction is insignificant compared to that of the forward reaction (*pseudo*-first order kinetics). To calculate the rate constant for such reactions, Equation 2.5 can be manipulated to produce an equation that describes a straight line:

$$\ln[A] - \ln[A_o] = -kt \tag{2.8}$$

Thus, for a first order reaction a plot of $\ln[A]$ versus time should be linear, with a gradient corresponding to $-k$. This first order relationship can break down when one or more of the reactants is limited, or there is a build-up of product. In these cases, a term to describe the limiting concentration should be introduced. The simplest, and often most common example is where the rate of the reaction is second order overall, but first order with respect to each of the reactants and in this case the rate of the reaction can be described by:

$$-\frac{d[A]}{dt} = k[A][B] \tag{2.9}$$

which, on integration and manipulation gives:

$$\frac{1}{([A_o] - [B_o])}\ln\frac{[B_o]}{[A_o]}\frac{[A]}{[B]} = kt \tag{2.10}$$

Thus, a plot of $\ln[A]/[B]$ versus t will have gradient $([A_o] - [B_o])k$.

2.4.4 Catalysed Systems

In a catalysed system, the spontaneous reaction may also occur, but in many cases it is insignificant compared to the catalysed rate of reaction. The effect of a catalyst is to accelerate the rate of both the forward and reverse reaction, allowing equilibrium to be reached much more quickly. The concentration of the reactants is still important, as this affects the probability that the catalyst will interact with the reactants and trigger the reaction.

2.5 KINETICS OF BIPHASIC REACTIONS

In a biphasic system, the same rules as above apply, however, the rate of the reaction and the position of the equilibrium are determined by the concentration of the reactants and products in the phase where the reaction takes place, rather than their overall concentration in the system. Exactly where the reaction actually takes place is still a matter of debate, with two locations proposed, specifically, at the interfacial layer between the two phases (model 1) and in the bulk of the catalyst-containing phase (model 2), as shown in Figure 2.9.

In model 1, the concentration of reactants in phase 1 reaching the catalyst will depend on the surface effects, whereas in model 2 the reactants must transfer into phase 2 before the reaction can take place. Because of these differences it has been suggested that the two models can be distinguished experimentally, since, if the reaction takes place in the bulk solvent the rate will be increased by the effect of co-solvents and other solubility promoters, whereas surfactants and other surface active compounds will affect the rate of the reaction occurring at the solvent interface. However, it is worth noting that some surface active compounds may also increase the surface area between the two solvents, thereby increasing catalyst accessibility in model 1, but also facilitating mass transfer of the reactants to phase 2 in model 2, thereby increasing the rate of both systems.

As for a single phase system, the rate of the reaction is still dependent on the probability of reactants meeting and therefore on the concentration of the reagents. However, in the biphasic system, the critical concentration of these components is no longer their total concentration in the whole system but the concentration where the reaction takes place. This concentration will be dependent on a number of factors, and the most influential are the rate of diffusion of the reactants to the catalyst and the relative solubility of the reagents in each phase. These two factors are interdependent, and will be considered in turn.

2.5.1 The Concentration of Reactants in Each Phase is Affected by Diffusion

In a homogeneous system, the rate of diffusion (J) across any given distance (x) is proportional to the concentration gradient of the reagent ($d[A]/dx$) and

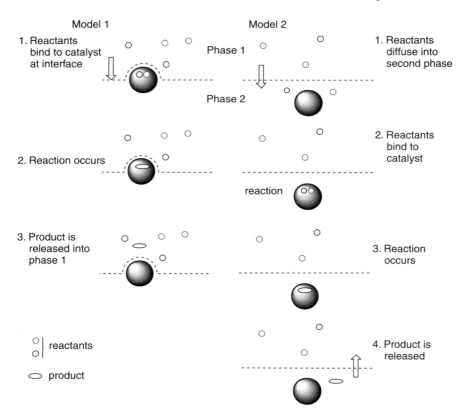

Figure 2.9 Schematic representation illustrating where the reaction takes place in biphasic reactions

these terms can be directly related by a diffusion coefficient D, and can be described by Equation 2.11. Like the rate Equation 2.4, Equation 2.11 describes an exponential curve, as shown in Figure 2.10.

$$J = D \frac{d[A]}{dx} \qquad (2.11)$$

For diffusion in a biphasic system, there is the additional complication of the phase boundary. Therefore, diffusion in each phase will be described by Equation 2.11, but in the region of the phase boundary different rules apply to take into account the mass transfer of the reactant from one phase to the other. Where the solubility of the solute is the same in both phases, the rate of diffusion across the phase boundary J for a solute moving from the higher concentration $[A]_1$ to the lower concentration $[A]_2$ through a film of thickness l is given by Equation 2.12, which also describes an exponential decrease in concentration, but

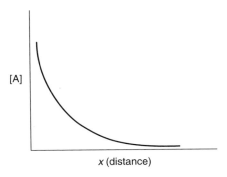

Figure 2.10   Diffusion in free solution

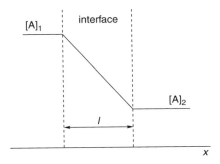

Figure 2.11   Diffusion across the phase boundary

as the film thickness l is usually very small in practice only a small fraction of the exponential curve will be considered, as shown by the linear representation in Figure 2.11.

$$J = \frac{D}{l}\frac{d([A]_1 - [A]_2)}{dx} \tag{2.12}$$

2.5.2   The Concentration of the Reactants and Products in the Reacting Phase is Determined by Their Partition Coefficients

In most biphasic systems, the solubility of the solute differs between the two phases. In this case, it is not the absolute concentration of the reagent that affects the rate of diffusion, but the concentration relative to the saturation of the solution. In the extreme example shown in Figure 2.12, although the actual concentration of solute A is higher in phase 1 than in phase 2, diffusion will proceed in the direction of phase 1 from phase 2, because phase 1 is less saturated by solute A than phase 2. The saturation is determined by the solubility of the solute in

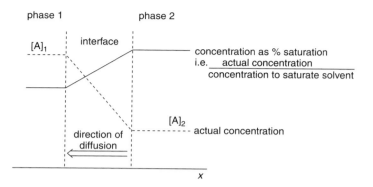

Figure 2.12 Diffusion over a phase boundary where the solubility in each phase differs

each solvent, and in this example solute A is much more soluble in phase 1 than in phase 2. Thus, more solute A can be dissolved in phase 1 before saturation is reached and therefore, even when the concentration of solute A is higher in phase 1 than in phase 2, the percentage saturation is less, driving diffusion from the low concentration in phase 2 towards the higher concentration in phase 1.

The distribution of solutes between the two phases can be estimated by measurement of the relative solubility of the reagent in each phase of the system under given physical conditions. Water and 1-octanol have become a standard solvent mixture for this [21], although other pairs of solvents may be used to produce, for example, a scale of preference for fluorous versus organic solvents, as described below. The quantity P, known as the *partition coefficient*, is defined as

$$P = \frac{[X]_{\text{solvent A}}}{[X]_{\text{solvent B}}} \tag{2.13}$$

where X is the substance whose partitioning is under investigation. Since P varies by at least 8 orders of magnitude, it is better expressed as $\log_{10}P$. Values of $\log_{10}P$ for some representative solvents are given in Table 2.2. This is an informative scale as it gives some indication of the affinity for water hydrophilicity compared to an organic phase organophilicity. This is clearly beneficial when designing chemical reactions in biphasic solvent systems. Some examples of the partitioning coefficients for different solutes in a fluorous biphasic system, using perfluoromethylcyclohexane and toluene as the two solvents are given in Table 2.3 [22].

2.5.3 The Partition Coefficients of the Reactants and Products May Alter the Position of the Equilibrium

As discussed earlier, a chemical reaction may occur in both directions. When the rate of the forward reaction is equal to that of the reverse the system is

Table 2.2 1-Octanol/water partition coefficients ($\log_{10}P$) for selected solvents at $25\,^\circ C^a$

Solvent	$\log_{10}P$	Estimated uncertainty
Dimethyl sulfoxide	−1.35	±0.20
N,N-dimethylformamide	−1.01	±0.20
Methanol	−0.74	±0.07
Acetonitrile	−0.34	±0.05
Ethanol	−0.30	±0.03
Acetone	−0.24	±0.10
Tetrahydrofuran	0.46	±0.10
Ethyl acetate	0.73	±0.05
Diethyl ether	0.89	±0.10
Dichloromethane	1.25	±0.15
Chloroform	1.97	±0.15
Benzene	2.13	±0.10
Toluene	2.73	±0.25
Carbon tetrachloride	2.83	±0.25
Cyclohexane^b	3.44	–
n-Hexane	4.00	±0.25

^{a} Values from Sangster J. *J. Phys. Chem. Ref. Data* 1989, **18**, 1111, which compares data from numerous sources for 611 organic compounds, except
^{b} from *CRC Handbook of Chemistry and Physics*, 80th Edition, CRC Press, Florida, 2000.

said to have reached equilibrium, i.e. there is no net change in the concentration of the reactants and the products although the reaction continues to take place. Equation 2.14 describes the equilibrium constant for the reaction where reactants A and B produce products X and Y and shows that this is dependent on the relative concentrations of the reactants and the products.

$$K_{eq} = \frac{[X][Y]}{[A][B]} \qquad (2.14)$$

As for the rate of diffusion, the equilibrium constant for a reaction in a biphasic system is not determined by the overall concentration of each reagent, but by their concentrations in the reaction phase. In some cases this can drive the forward reaction to completion, and in other cases it can be inhibitory, depending on the relative concentrations of the reactants and products. In model 1, where the reaction takes place at the phase boundary, the effective concentration of the reactants and products will be that in phase 1, and assuming each has an equivalent solubility, the equilibrium position will approach that of a homogeneous system. Where the reaction takes place in the bulk solvent, as in model 2, the equilibrium position is very much dependent on the solubility of the reagents in phase 2. For example, if the product is less soluble in phase 2 than the reactant, as the product is formed it will diffuse back into phase 1, reducing its concentration in phase 2 where the reaction is occurring and therefore the reaction will

Table 2.3 Partition coefficients at 24 °C for different solutes in the $CF_3C_6F_{11}/CH_3C_6H_5$ biphasic system

Compound	P	$\log_{10}P$
	0.98	0.01
	10.36	1.02
	9.75	0.98
	10.24	1.01
	1.28	0.11
	0.06	−1.24
	0.04	−1.36
	0.04	−1.45
	0.02	−1.6
	0.02	−1.7
	0.01	−1.92
	0.009	−2.04
	0.29	−0.54
	0.39	−0.41

continue in the forward direction. In special cases, where the product is poorly soluble in the reacting phase, the reaction is driven in the forward direction so that it can go to completion. The equilibrium position in this latter model is also dependent on the rate of the reaction relative to the rate of diffusion. In fast reactions, concentration gradients of the reactant and product form through phase 2 (see below), thereby affecting the equilibrium position. So, away from the phase boundary, the product may accumulate at the expense of the reactant, favouring reverse reaction.

2.5.4   Effect of Diffusion on Rate

In a homogeneous system, the rate of diffusion in the system can be directly related to the rate of the reaction as it governs the number of times the catalyst will interact with the reactants over a set time. In a biphasic system, diffusion still affects the rate of reaction, as this is dependent on the catalyst and reactants meeting. However, the rate of diffusion also affects the time it takes for the reactants to reach the place where the reaction takes place. How diffusion affects rate depends on the catalytic turnover.

For a slow reaction occurring in the bulk liquid of phase 2, phase transfer and diffusion may not be rate limiting. Here, the reactants will enter phase 2 more quickly than they are consumed by the reaction, and thus become distributed at equal concentration throughout this phase (see Figure 2.13a). In this case the rate of the reaction in phase 2 will be described by Equation 2.3, with the rate of reaction being directly proportional to the concentration of the reactants in phase 2 and uniform throughout the phase. Similarly, for a slow reaction occurring at the phase boundary, the rate of diffusion of the reactants will exceed their

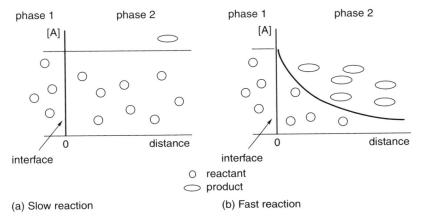

Figure 2.13   Variation of the concentration of reactants with distance from the solvent interface in bulk liquid

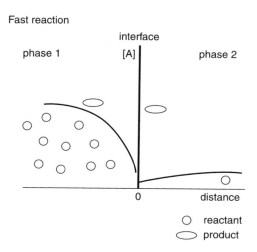

Figure 2.14 Variation of concentration of reactants with distance from the solvent–solvent interface in a reaction limited by diffusion

consumption, thus they will be continually replenished and the concentration of the reagents will be distributed uniformly throughout each phase.

In contrast, a fast reaction rate will result in steep concentration gradients for the reactants and a higher reaction rate near the solvent interface. This concept is represented diagrammatically in Figure 2.13b, where the concentration of reactant A is almost as high as that in phase 1 at the solvent interface, but plummets as it is rapidly consumed by the reaction. Thus, for a fast reaction, the majority of reactant is converted to product near the phase boundary layer and the rate of the reaction is limited by the rate of phase transfer and diffusion.

A similar prediction can be made for the concentration distribution of reagents for a diffusion limited reaction occurring at the phase boundary. The concentration of the reactants decreases around the phase boundary, as this is the site where they are consumed. In Figure 2.14, it is assumed that the reactant A has about one tenth of the solubility in phase 2 compared to phase 1, thus in most cases some of reactant A will diffuse across the phase boundary into this phase. As in phase 1, the concentration distribution will not be equal throughout the phase, but it will be lower in proximity to the phase boundary. If the reaction is very fast, reactant A will be consumed at the phase boundary and will therefore not enter phase 2.

2.5.5 Determining the Rate of a Reaction in a Biphasic System

As the measured rate of the reaction in a biphasic system can be dramatically affected by diffusion, the measured rate of reaction is termed the observed rate of reaction, which is equivalent to the actual rate of the reaction including a factor which describes the rate of diffusion. For a slow reaction, where diffusion is not rate limiting and therefore does not affect the rate, then this factor will be equal to 1 and the rate equation is equivalent to Equation 2.3. However, for a fast reaction, the rate will be limited by diffusion as this determines the maximum rate at which the reactants can be replenished during the catalytic cycle. In this case, the rate of the reaction is limited by the concentration of the reactant and therefore is limited by the rate of its diffusion across the phase boundary and decreases with distance from the phase boundary (see Figure 2.13b). There are different ways in which the rate can be altered, including improved mixing of the two phases and the addition of phase transfer catalysts.

In many biphasic systems, constant stirring creates a fine emulsion, where droplets of one solvent become suspended in the other. In this emulsion the surface area between the two phases is increased, providing a bigger surface for either the catalytic reaction to occur or the reactants to diffuse across to react in the bulk solvent.

Phase transfer catalysts can be used to increase the solubility of reactants in the phase where the reaction takes place. Usually these catalysts are organophilic salts that pair with anionic reactants to increase their solubility in organic solvents. Phase transfer catalysis is described in more detail in Chapter 5.

2.6 CONCLUSIONS

There are many good reasons for using multiphasic methods to carry out reactions that employ homogeneous catalysts. In general, the catalyst is immobilized in one phase and the reactants and products are supported in other phases. Ideally, once the reaction is complete the products can be removed without contamination by the catalyst and the catalyst phase is ready for immediate reuse. However, a perfect system is extremely difficult to obtain in practice, although many solvent combinations and catalyst design strategies have been developed which come close to the ideal situation.

REFERENCES

1. Manassen J. In *Catalysis: Progress in Research*, Bassolo F. and Burwell R. L. (eds), Plenum Press, London, 1973, 183.
2. Horváth I. T. and Rábai J. *Science* 1994, **266**, 72.
3. Cornils B. and Herrmann W. A. *J. Mol. Catal. A Chemical* 1997, **116**, 27.
4. Ryu I., Matsubara H., Yasuda S., Nakamura H. and Curran D. P. *J. Am. Chem. Soc.* 2002, **124**, 12946.
5. Joó F. and Kathó Á. *J. Mol. Catal. A Chemical* 1997, **116**, 3.
6. Kohlpaintner C. W., Fischer R. W. and Cornils B. *Appl. Catal. A Gen.* 2001, **221**, 219.
7. Arts S. J. H. F., Momberg E. J. M., van Bekkum H. and Sheldon R. A. *Synthesis* 1997, **6**, 597.

8. Bhanage B. M. and Arai M. *Catal. Rev.* 2001, **43**, 315.
9. (a) Herndon J. W. *Coord. Chem. Rev.* 1999, **181**, 177; (b) Cornils B. *J. Mol. Catal. A Chemical* 1999, **143**, 1.
10. Mecking S. Held A. and Bauers F. M. *Angew. Chem., Int. Ed. Engl.* 2002, **41**, 544.
11. Horváth I. T. *Acc. Chem. Res.* 1998, **31**, 641.
12. Li H. M., Shu H. M., Ye X. K. and Wu Y. *Prog. Chem.* 2001, **13**, 461.
13. Dupont J., de Souza R. F. and Suarez P. A. Z. *Chem. Rev.* 2002, **102**, 3667.
14. Dullius J. E. L., Suarez P. A. Z., Einloft S., de Souza R. F., Dupont J., Fischer J. and De Cian A. *Organometallics* 1998, **17**, 815.
15. Dyson P. J., Ellis D. J. and Welton T. *Can. J. Chem.* 2001, **79**, 705.
16. Sellin M. F., Webb P. B. and Cole-Hamilton D. J. *Chem. Commun.* 2001, 781.
17. Quici S., Cavazzini M., Ceragioli S., Montanari F. and Pozzi G. *Tetrahedron. Lett.* 1999, **40**, 3647.
18. Carmichael A. J., Earle M. J., Holbrey J. D., McCormac P. B. and Seddon K. R. *Org. Lett.* 1999, **1**, 997.
19. Zamaraev K. I. *Top. Catal.* 1996, **3**, 1.
20. Tanaka K. and Toda F. *Chem. Rev.* 2000, **100**, 1025.
21. (a) Sangster J. *J. Phys. Chem. Ref. Data* 1989, **18**, 1111; (b) Marrero J. and Gani R. *Ind. Eng. Chem. Res.*, 2002, **41**, 6623.
22. Rocaboy C., Rutherford D., Bennett B. L. and Gladysz J. A. *J. Phys. Org. Chem.* 2000, **13**, 596.

3   Reactions in Fluorous Media

The term 'fluorous' was coined as an analogy to 'aqueous' for highly fluorinated alkanes, ethers and tertiary amines [1]. These compounds differ markedly from the corresponding hydrocarbon compounds to the extent that such compounds commonly give bilayers with conventional organic solvents. In this chapter, we will discuss the different approaches towards carrying out reactions in fluorous media and describe how reactants and catalysts can be engineered to be preferentially soluble in fluorous solvents.

3.1   INTRODUCTION

Fluorous compounds with appropriate melting and boiling points can be used as solvents for reactions. Fluorous compounds that have been used as solvents for a variety of reactions include perfluoromethylcyclohexane (PP2) and perfluorohexane (Figure 3.1).

Since fluorous solvents tend to mix poorly with common organic solvents, they have been widely studied in biphasic catalysis where a reagent or catalyst is anchored in the fluorous phase and separated from the organic phase at the end of a reaction. However, unlike water (see Chapter 5), in addition to forming bilayers with conventional solvents, some combinations of organic and fluorous solvents demonstrate increased miscibility at elevated temperatures. In some cases, heating can even result in a completely homogeneous mixture. Accordingly, it is possible to carry out reactions under either heterogeneous *or* homogeneous conditions. So, for example, whilst PP2 and chloroform form a biphasic system at room temperature, they form a single phase at 50.1 °C [2]. Similarly, toluene and perfluorodimethylcyclohexane (PP3) are immiscible at room temperature but miscible at 70 °C, as shown in Figure 3.2. By careful design of the catalysts and/or reagents, a reaction can be carried out homogeneously with the added advantage of a biphasic separation of catalyst and products after the reaction is complete. This technique is particularly applicable to those reactions where a nonpolar compound is converted to products with a higher polarity since the more polar the compound, the lower the solubility in the fluorous phase.

Chemistry in Alternative Reaction Media   D. Adams, P. Dyson and S. Tavener
© 2004 John Wiley & Sons, Ltd   ISBNs: 0-471-49848-3 (Cloth); 0-471-49849-1 (Paper)

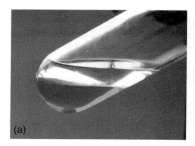

Perfluoromethylcyclohexane, PP2   Perfluorohexane, FC-72

Figure 3.1   Examples of fluorous solvents

Figure 3.2 (Plate 4)   (a) A biphase of toluene (top) and a fluorous-soluble catalyst in PP3 (bottom) at 25 °C; (b) the same mixture when heated to 70 °C. (Photographs by James Sherrington)

3.2   PROPERTIES OF PERFLUORINATED SOLVENTS

Perfluorinated alkanes and cycloalkanes are prepared from the corresponding hydrocarbons, either by electrochemical fluorination or by cobalt trifluoride fluorination [3]. Many perfluorinated solvents are available commercially covering a wide selection of boiling points and densities. Some examples of perfluorinated solvents are listed in Table 3.1 together with their key physical properties.

Although there are obvious variations within these compounds, they share many common features. Perfluorinated compounds have high densities compared to their hydrocarbon analogues, generally between 1.7 and $1.9\,\mathrm{g\,cm^{-3}}$. They have very low polarities and generally have low solubilities in both water and most organic solvents. Because of the low polarizability of the electrons in C–F bonds and the low availability of the lone-pair electrons of fluorine, fluorocarbons exhibit very weak van der Waals interactions. As a result, gases generally have very high solubilities in fluorocarbons since it costs very little energy to replace one perfluorocarbon molecule with a different molecule which also has little interaction with its neighbours. The high solubility of oxygen in perfluorocarbons has resulted in their use as artificial blood substitutes [5]. Kamlet–Taft α, β and π^* parameters have been obtained for some perfluorinated solvents (Table 3.2). That these solvents are not HBDs and are extremely poor HBAs is reflected in the very low α values and negative β values. The strongly negative π^* indicates that

Table 3.1 Selected data for some perfluorinated compounds

Perfluorocarbon	Formula	Mp (°C)	Bp (°C)	$P_s{}^a$	O_2 sol[b]	Density (g cm^{-3})
n-Perfluorooctane	C_8F_{18}	−25	103–105	0.55	52.1	1.74
n-Perfluorohexane, FC-72	C_6F_{14}	−87	57	0.00	–	1.68
Perfluoromethyl-cyclohexane, PP2	C_7F_{14}	−44.7	76.1	0.46	–	1.79
Perfluorodimethyl-cyclohexane, PP3	C_8F_{16}	−55	101–102	0.58	–	1.83
Perfluorotributyl amine	$(C_4F_9)_3N$	−50	178–180	0.68	38.4	1.90
Perfluorooctyl bromide	$C_8F_{17}Br$	–	142	–	52.7	1.89
FC-75[c]	$C_8F_{16}O$	–	102	–	52.2	1.78
Perfluorodecalin	$C_{10}F_{18}$	−10	141	–	40.3	1.95
D80	[d]	−79[f]	84	–	45[g]	1.73
D100	[e]	−104[f]	104	–	41[g]	1.77

[a] Spectral polarity index based on an α-perfluoroalky-β,β-dicyanovinyl dye (see Chapter 1), the higher the number, the more polar the compound. For comparison, n-hexane has a P_s of 2.56, benzene 6.95 [4].
[b] ml O_2/100 ml solvent.
[c] Mainly perfluorobutyltetrahydrofuran.
[d] General formula $CF_3[(O(CF_3)CFCF_2)_m(OCF_2)_n]OCF_3$, average molecular weight 390.
[e] General formula $CF_3[(O(CF_3)CFCF_2)_m(OCF_2)_n]OCF_3$, average molecular weight 416.
[f] Pour point.
[g] Solubility of air, ml air/100 ml solvent.

Table 3.2 Kamlet–Taft parameters for three perfluorinated solvents [6]

Solvent	α	β	π^*
n-Perfluorohexane	0.00	−0.08	−0.41
Perfluorodecalin	0.00	−0.05	−0.32
Perfluoromethylcyclohexane	0.00	−0.06	−0.40

these solvents are extremely difficult to polarize, which may explain their tendency to form biphases with organic solvents. Attractions between hydrocarbons formed by dispersive forces (i.e. temporary dipoles which induce dipoles in their neighbours by polarization) outweigh those formed between the hydrocarbon and the hard-to-polarize fluorocarbon.

Perfluorocarbons are chemically inert, are nonflammable and have low toxicities. Unlike the halofluorocarbons, perfluorocarbons are not ozone-depleting and have been used as CFC replacements [7]. However, perfluorocarbons have very long lifetimes in the atmosphere (>2000 years) [8] and this has consequences for their use as solvents since any process has to take into account the fact that any loss of solvent to the atmosphere is extremely undesirable.

3.3 DESIGNING MOLECULES FOR FLUOROUS COMPATIBILITY

For catalysts or reagents to have a high affinity for a fluorous phase it is often necessary for them to be derivatized with long perfluoroalkyl chains such as C_6F_{13} or C_8F_{17}. These are known as 'fluorous ponytails', and work on the basis that 'like-dissolves-like'. This frequently results in molecules which have three domains: a fluorous domain which controls the solubility, an organic domain which 'shields' the reactive centre from the electron-withdrawing effect of the perfluoroalkyl group, and lastly the functional group which dictates the reactivity of the molecule, as shown in Figure 3.3. It is often quoted that the percentage fluorine by weight within the molecule needs to be greater than 60 % to confer preferential solubility in a fluorous solvent [9].

Often, an organic group is placed between the perfluoroalkyl group and the remainder of the compound in an attempt to minimize the electron-withdrawing effect of the fluorinated group (fluorine is the most electronegative element). Without this insulating spacer group, the electron density on the reactive group would be seriously reduced. For a phosphine for example, the electron density might become reduced to the point where it would be unable to act as an effective ligand. Suitable spacer groups include a number of CH_2 units, OCH_2 and $Si(Me)_2$. Aryl spacers have also been used, either alone or in combination with the other spacer groups. Whilst preparation of the above compounds may seem trivial, this is often not the case, although the range of reported reactions and available building blocks is ever increasing. Some examples of compounds that have been synthesized for use in a fluorous biphase are shown in Figure 3.4.

Because of the current high cost of perfluorinated compounds and solvents, it really only makes sense to use them for reactions which produce expensive products or where the catalyst is expensive and recycling and reuse is essential. There are many examples of transition metal catalysed reactions that fall into this category. In many cases, such catalysts utilize modifying ligands such as phosphines and phosphites and it is of no surprise that a substantial amount of work has been carried out preparing analogues of known ligands derivatized with perfluoroalkyl

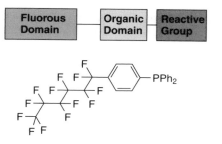

Figure 3.3 The different domains of a fluorous molecule

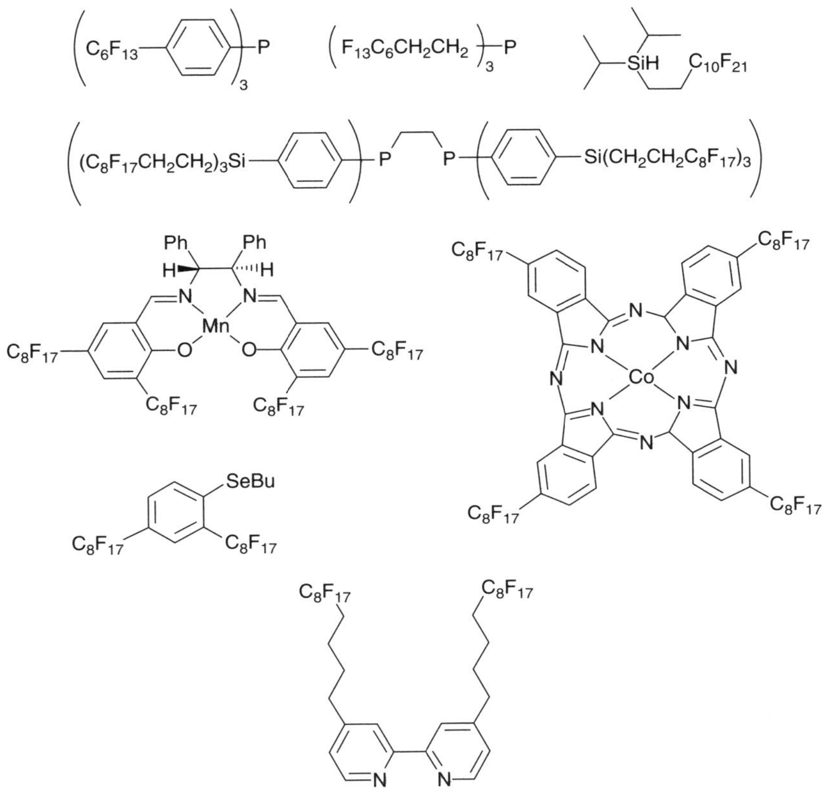

Figure 3.4 Some examples of perfluoroalkyl-substituted molecules suitable for reactions in fluorous solvents

groups. Many different methodologies now exist for the derivatization of organic compounds and the most important of these methods are described below.

(i)   Simple alkyl phosphines can be formed via reaction of PH₃ with a perfluoroalkyl-functionalized alkene [10] catalysed by 2,2′-azobis(isobutyronitrile) (AIBN) or by reaction of PCl₃ with the Grignard reagent derived from a perfluoroalkyl-functionalized iodide [11]. Phosphites can be formed via direct reaction of a fluorinated alcohol with a chlorophosphine in the presence of a base to neutralize the HCl formed [12] (Scheme 3.1). These reactions result in alkyl phosphines and phosphites that incorporate a number of CH₂ spacer units.

(ii) Aryl spacer units can be incorporated via copper mediated coupling reactions [13]. Here, perfluoroalkyliodides react with iodoaromatic compounds in the presence of copper in a dipolar aprotic solvent, such as dimethyl

$$ICH_2CH_2C_6F_{13} \xrightarrow[\text{(ii) } PCl_3]{\text{(i) Mg, Et}_2O} P(CH_2CH_2C_6F_{13})_3$$

$$HOCH_2CH_2C_6F_{13} \xrightarrow[\text{Et}_2O]{PCl_3, \text{ Et}_3N} P(OCH_2CH_2C_6F_{13})_3$$

$$PH_3 + 3CH_2CHC_6F_{13} \xrightarrow{\text{AIBN}} P(CH_2CH_2C_6F_{13})_3$$

Scheme 3.1

sulfoxide (DMSO). 2, 2′-Bipyridyl (bipy) is sometimes included to lower the reaction temperature and co-solvents such as hexafluorobenzene can also be used to aid solubility [14]. Such reactions allow the direct attachment to an aromatic ring, which can be further reacted to give a derivatized phosphine or phosphite as shown in Scheme 3.2. Utilizing such a methodology, a range of perfluorinated phosphines, phosphinites, phosphonites and phosphates have been prepared [14]. Indeed, a simple analogue of triphenylphosphine, tris(4-tridecafluorohexylphenyl)phosphine, is now commercially available [15].

(iii)   Aryl and CH_2 spacer units can be combined using a number of methodologies. For example, the Grignard reagent derived from a *1H,1H,2H,2H*-perfluoroalkyl iodide can be reacted with an aryl halide. However, yields are low with the magnesium reagent [16], but the use of an organozinc reagent increases the yield [17]. An alternative is to carry out a Heck coupling reaction between a *1H,1H,2H*-perfluoro alkene and an aryl halide or arenediazonium salt [18]. This results in an aromatic with an ethylene-spaced perfluoroalkyl group. This can then be hydrogenated to give the required alkyl spacer (Scheme 3.3).

These types of compounds can also be formed via a Wittig reaction [19]. Triphenylphosphine can be quaternized with a suitable alkyl iodide, and the resultant perfluoroalkylated phosphonium salt will react with aldehydes to give fluorinated alkenes which are easily hydrogenated (Scheme 3.4). This methodology has recently been expanded to the formation of perfluoroalkylated pyridines [20].

(iv)   An alternative spacer unit which may be combined with the aryl group is OCH_2 [21]. Insertion of this spacer group can be achieved by reaction of a phenol with a perfluoroalkyl nonafluorobutane sulfonate in the

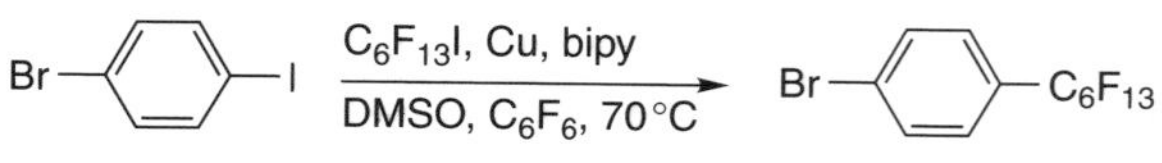

$$Br\!-\!\!\left\langle\underset{}{\bigcirc}\right\rangle\!\!-\!I \xrightarrow[\text{DMSO, } C_6F_6, 70\,°C]{C_6F_{13}I, \text{ Cu, bipy}} Br\!-\!\!\left\langle\underset{}{\bigcirc}\right\rangle\!\!-\!C_6F_{13}$$

Scheme 3.2

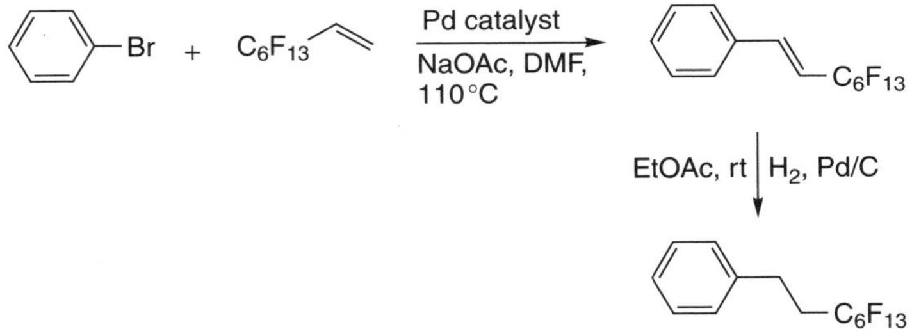

Scheme 3.3

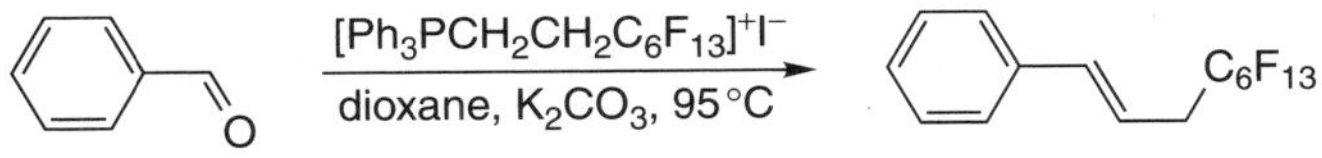

Scheme 3.4

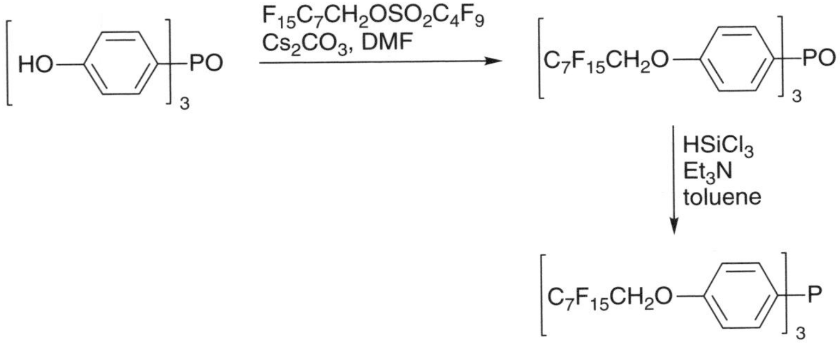

Scheme 3.5

presence of a base (Scheme 3.5). A tris-substituted phosphine oxide have been prepared by attaching the fluorous ponytail to *tris*−4-hydroxyphenyl phosphine oxide, before reduction to yield the phosphine.

(v) Silyl spacer units have also has been combined with aryl spacer groups [22]. Such compounds are formed via bromination of a perfluoroalkyl deriva-tized silane followed by reaction with an aryl Grignard (Scheme 3.6). One major advantage of this route is that it allows up to three perfluorinated ponytails to be attached to each aromatic ring.

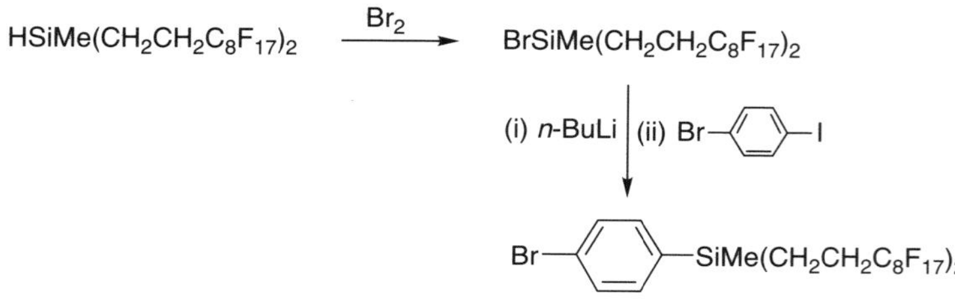

$$HSiMe(CH_2CH_2C_8F_{17})_2 \xrightarrow{\quad Br_2 \quad} BrSiMe(CH_2CH_2C_8F_{17})_2$$

(i) n-BuLi $\bigg|$ (ii) Br—⟨⟩—I

Br—⟨⟩—SiMe(CH_2CH_2C_8F_{17})_2

Scheme 3.6

3.4 PROBING THE EFFECT OF PERFLUOROALKYLATION ON LIGAND PROPERTIES

The lone pair of electrons on the phosphorus atom of a phosphorus(III) ligand are available for donation to a metal centre (a σ bond). This bonding is enhanced by so-called back-bonding from the filled d-orbitals of the metal into empty σ^*-orbitals of the phosphorus(III) ligand. This gives the bond a degree of π-character and strengthens the bond. Electron-withdrawing groups on the phosphorus ligand will pull electron density away from the phosphorus centre and so decrease the amount of σ-donation possible. The spectroscopic and structural properties of transition metal complexes have been used to assess the electronic and steric impact of perfluoroalkyl substituents on the donor properties of such ligands. Since many transition metal catalysts are generated *in situ* from the free ligand and metal-containing precursors, the successful development of the fluorous biphase approach to catalysis depends upon the reactivities of these ligands mirroring those of the unmodified phosphine ligands. The reactions of perfluoroalkylated phosphine ligands with a range of platinum metal reagents (Scheme 3.7) have been investigated in detail. Information on the effect of adding perfluoroalkyl groups to triphenylphosphine derivatives was deduced spectroscopically by comparing the $^1J_{Pt\text{-}P}$ coupling constants of the *cis*-[PtCl$_2$L$_2$] complexes of the fluorous phosphines with the hydrocarbon-parent ligands as shown in Table 3.3 [23].

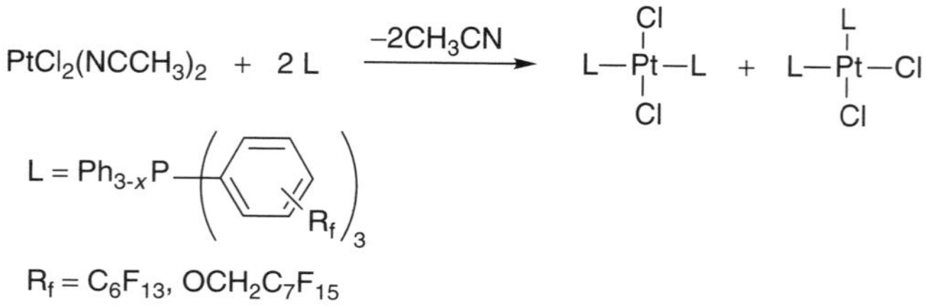

$$PtCl_2(NCCH_3)_2 + 2\,L \xrightarrow{\quad -2CH_3CN \quad} \begin{array}{c} Cl \\ | \\ L-Pt-L \\ | \\ Cl \end{array} + \begin{array}{c} L \\ | \\ L-Pt-Cl \\ | \\ Cl \end{array}$$

$$L = Ph_{3-x}P\left(\!\!\left\langle\!\!\left\langle \right\rangle\!\!\right\rangle_{\!\!R_f}\right)_{\!\!3}$$

$$R_f = C_6F_{13},\ OCH_2C_7F_{15}$$

Scheme 3.7

Table 3.3 Comparison of NMR ($^1J_{Pt-P}$) and IR (υ_{CO}) for metal complexes derived from both fluorinated and nonfluorinated phosphines

Phosphine	$^1J_{Pt-P}$	Rh $\upsilon_{CO}(cm^{-1})$	Ir $\upsilon_{CO}(cm^{-1})$
PPh$_3$	3672	1965	1953
PPh$_2$(4-C$_6$H$_4$C$_6$F$_{13}$)	3653	1982	1959
PPh$_2$(3-C$_6$H$_4$C$_6$F$_{13}$)	3633	1980	–
PPh$_2$(2-C$_6$H$_4$C$_6$F$_{13}$)	–	1965	–
PPh(4-C$_6$H$_4$C$_6$F$_{13}$)$_2$	3635	1983	1972
PPh(3-C$_6$H$_4$C$_6$F$_{13}$)$_2$	3602	1984	–
P(4-C$_6$H$_4$C$_6$F$_{13}$)$_3$	3615	1993	1979
P(3-C$_6$H$_4$C$_6$F$_{13}$)$_3$	–	1992	–
P(4-C$_6$H$_4$OCH$_2$C$_7$F$_{17}$)$_3$	3680	1977	1967

These investigations indicated that, for the most part, the strongly electron-withdrawing perfluoroalkyl substituents do make the phosphines weaker σ-donors. The $^1J_{Pt-P}$ coupling constant data decreases with the number of perfluoroalkyl substituents, reflecting the reduction in the σ-donor power of the phosphorous atoms. However, perfluoroalkylation does not reduce the electron density to the extent that these compounds are ineffective as ligands. On the other hand, the size of the perfluoroalkyl groups can have steric implications such that, for the *ortho*-substituted phosphine, the thermodynamically less favoured *trans*-[PtCl$_2$L$_2$] is favoured in preference to the *cis* isomer. In addition, the π-bonding characteristics of the ligands were investigated by measuring the infrared carbonyl stretching frequencies, υ_{CO}, of the complexes *trans*-[MCl(CO)L$_2$] (M = Rh, Ir) and again show the influence of the strongly electron-withdrawing perfluoroalkyl group. Together, the data indicate that *meta* and *para* substitution reduces the σ-donor properties and increases the π-acceptor behaviour of the phosphine ligands by a similar amount. For the *ortho*-derivatized ligand, the υ_{CO} data for *trans*-[RhCl(CO)(PPh$_2$(2-C$_6$H$_4$C$_6$F$_{13}$))] are very similar to the parent *trans*-[RhCl(CO)(PPh$_3$)], indicating that the decrease in basicity is cancelled out by the increase in cone angle. Similar comparisons have been made for the perfluoroalkyl derivatized phosphites [24].

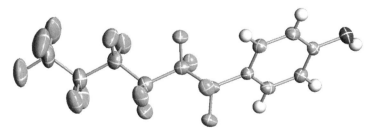

Figure 3.5 (Plate 5) Crystal structure of 4-tridecafluorohexylphenol [24]

The steric bulk of the perfluoroalkyl group can be demonstrated by examining crystal structures of suitable compounds. For example, the crystal structure of the *para*-substituted phenol clearly shows the size of the C_6F_{13} group with respect to the aromatic ring (Figure 3.5).

3.5 PARTITION COEFFICIENTS

Optimization of a fluorous-organic biphasic system requires knowledge of the solubility of the fluorous catalysts and reagents in the fluorous solvents with

Table 3.4 Partition coefficients for selected fluorous compounds in the $CF_3C_6F_{11}/CH_3C_6H_5$ biphase system [25]

Compound	P	$\log_{10}P$
—(CH$_2$)$_3$C$_8$F$_{17}$	0.980	−0.008
—(CH$_2$)$_3$C$_6$F$_{13}$ / (CH$_2$)$_3$C$_6$F$_{13}$	2.802	0.447
—(CH$_2$)$_3$C$_8$F$_{17}$ / (CH$_2$)$_3$C$_8$F$_{17}$	10.36	1.101
P((CH$_2$)$_3$C$_6$F$_{13}$)$_2$	3.648	0.562
P((CH$_2$)$_3$C$_8$F$_{17}$)$_2$	14.87	1.172
P(—C$_6$H$_4$—C$_6$F$_{13}$)$_3$	4.405	0.644
P(—C$_6$H$_4$—(CH$_2$)$_2$C$_6$F$_{13}$)$_3$	0.901	−0.045
P(—C$_6$H$_4$—(CH$_2$)$_3$C$_6$F$_{13}$)$_3$	0.266	−0.058
P(—C$_6$H$_4$—(CH$_2$)$_3$C$_8$F$_{17}$)$_3$	2.145	0.331

respect to organic solvents. One way of measuring the respective solubilities is by establishing the partition coefficient. Here, the species in question is dissolved in a mixture of an equal amount of a fluorous solvent and an organic solvent. The biphasic system is stirred, then allowed to come to equilibrium and then the amount of the species in each phase is determined. This is done typically gravimetrically, by gas chromatography or by atomic absorption. Some examples of measured partition coefficients are shown in Table 3.4 (values for other compounds are shown in Chapter 2).

Study of the factors that affect partition coefficients allows the rational design of subsequent generations of catalysts and reagents. It is clear from the data that, for simple aromatic compounds, the solubility in fluorous solvents increases with increasing number of fluorophilic ponytails. The solubility is also improved with increasing length of the perfluoroalkyl group. The increase in partition coefficient with increasing length of perfluoroalkyl tail is also observed for other compounds, for example the menthyl substituted phosphines. For the triaryl phosphines, the presence of spacer groups between the aromatic ring and the ponytail decreases the partition coefficient, but, again, the partition coefficient increases with increasing length of the tail. However, it is important to note that the partition coefficients are simply a measure of the relative solubility of the compound in the two solvents quoted and is not a measure of the absolute solubility. Indeed, at higher chain lengths, absolute solubilities can decrease, despite the partition coefficient increasing!

3.6 LIQUID–LIQUID EXTRACTIONS

When reactions are carried out in a fluorous phase or for that matter in any biphasic system, the products can often be recovered by simple phase separation. If required, the fluorous phase can be washed with further organic solvent to recover any residual product that remains in the fluorous phase. However, in catalysed reactions, efficient recycling of the catalyst is critical to the success of the reaction. The cost of derivatizing the modifying ligands means that any unrecovered catalyst has serious implications for the economics of the process.

Most simply, catalyst recovery is achieved by pouring off the upper, less dense phase, and the fluorous phase may be reused in another reaction. For example, long chain alkenes may be hydroformylated in a fluorous biphase system using a rhodium metal catalyst with either $P(CH_2CH_2C_6F_{13})_3$ or $P(C_6H_4C_6F_{13})_3$ as ligands (see Chapter 8) [26]. With the trialkyl phosphine, a semi-continuous process was used. After each reaction, the mixture was allowed to cool and then the organic layer removed. The fluorous phase was kept in the reactor and reused and, over nine consecutive runs, a total turnover number (TON, i.e. moles of product formed per mol of catalyst) of greater than 35 000 was measured with only a 4.2 % loss of rhodium. This value is equivalent to 1.18 ppm of rhodium lost for each mol of product. The loss of rhodium was ascribed to the low, but

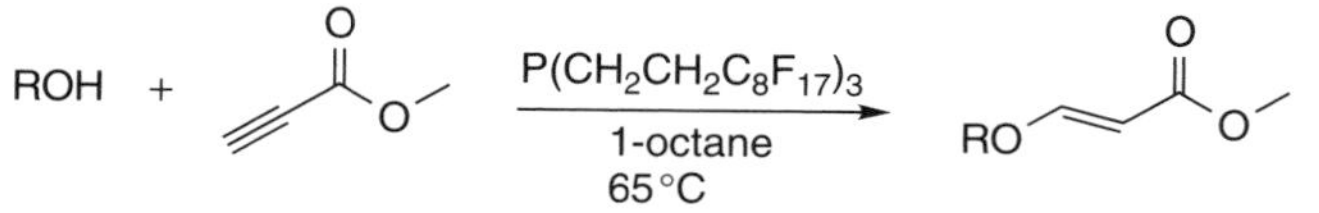

$$ROH + \underset{\text{(methyl propiolate)}}{\overset{O}{\underset{\|}{\bigg/\!\!\!\!\!\equiv\!\!\!-C-O-}}} \xrightarrow[\substack{\text{1-octane}\\65\,°C}]{P(CH_2CH_2C_8F_{17})_3} RO\overset{O}{\underset{\|}{-CH=CH-C-O-}}$$

Scheme 3.8

finite, solubility of the catalyst in the product phase. Unfortunately, the free ligand is also leached into the organic phase, which resulted in a decrease in reaction selectivity. With the triaryl phosphine, it is also possible to recover the product with only a very low level of leaching of the rhodium catalyst (in some cases as low as 0.05 %) [27].

An exciting development in fluorous biphasic chemistry is the demonstration that recycling can be achieved simply on the basis of the thermomorphic properties of the catalyst [28]. It was noted that appending a number of fluorous ponytails to a molecule often has the effect of lowering the melting point. For example, $P(C_2H_4C_8F_{17})_3$ melts at only 47 °C. These phosphines often show a marked increase in solubility in conventional solvents at elevated temperatures, especially near their melting points. When these phosphines were used to catalyse the reaction of alcohols with methyl propiolates, (Scheme 3.8) good yields were obtained in benzotrifluoride at room temperature. The catalyst could easily be recycled by conventional means using a benzotrifluoride/1-octane biphasic system. Interestingly, the phosphine showed a marked increase in solubility in 1-octane at higher temperature that may be exploited in the synthesis. The reaction can be carried out in 1-octane only at 65 °C and, when the reaction is complete, the solution is cooled which leads to precipitation of the catalyst. Simple decantation led to the successful separation of the phosphine, which was used for a further four runs without deterioration in activity.

3.7 SOLID SEPARATIONS

A problem with simple phase separation in fluorous biphasic systems is the cost associated with the high fluorine content required to anchor a catalyst exclusively in the fluorous phase. In an effort to overcome such difficulties, the use of *Fluorous Reverse Phase Silica Gel* (FRPSG) [29] has been developed as a useful tool in organofluorine chemistry. The method relies on the low retention of most organic moieties on fluorinated silica and is conceptually the same as using reverse phase silica gel for liquid chromatography. The separation of a fluorinated species from organic moieties is possible by eluting the mixture through FRPSG with an organic solvent. Due to their low solubility in organic solvents, the fluorinated species are eluted last, allowing facile separation of, for example, fluorinated catalysts from organic products [30] (Figure 3.6).

The crucial point is that derivatization with a perfluoroalkyl group is required, in this instance, only to confer a sufficient polarity difference between species

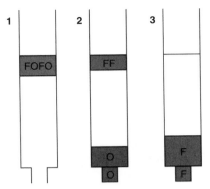

1. Load column with fluorous (F) and organic (O) compounds
2. Elute organic compounds with polar solvent
3. Elute fluorous compounds with nonpolar solvent

Figure 3.6 Separation of a fluorous compound from an organic compound using FRPSG

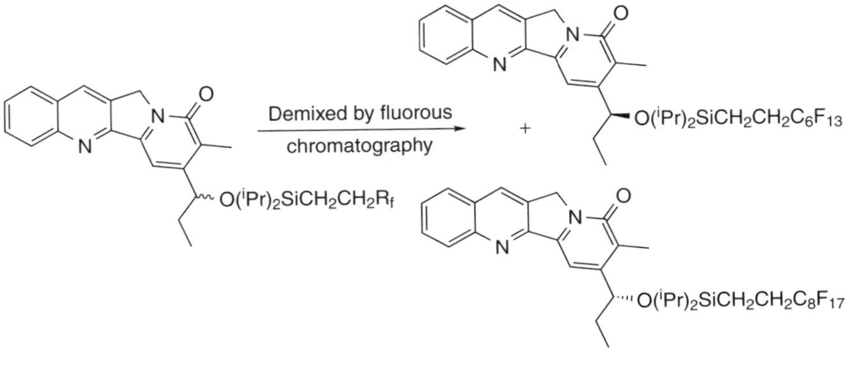

Scheme 3.9

such that separation on FRPSG is possible. A preferential solubility in a fluorous solvent is no longer required in contrast to the fluorous biphase reactions described above where preferential solubility in a fluorous solvent is crucial to the success of the entire process. The FRPSG technique has been applied to the separation of two different mappicines, the (R)-enantiomer tagged with a silyl protecting group bearing a C_6F_{13} chain, the (S)-enantiomer tagged with a C_8F_{17} chain as shown in Scheme 3.9 [31]. These were formed from tagged alcohols and an equimolar mixture of both carried through a sequence of reactions to produce the mappicines, which could then be separated by HPLC separation over FRPSG.

The separation and successful reuse of a metal catalyst using FRPSG has also been demonstrated [32]. A fluorinated nickel complex was found to catal-

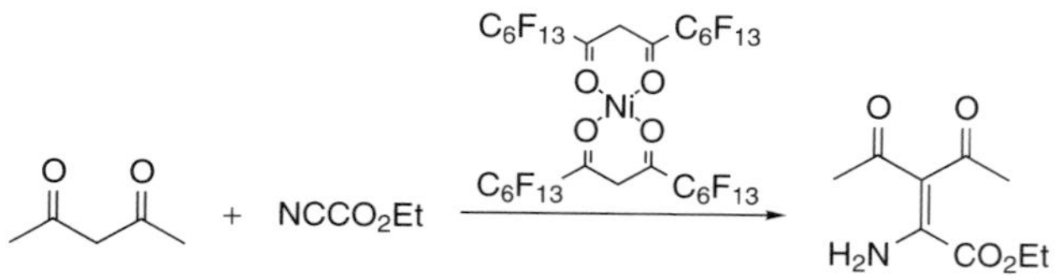

Scheme 3.10

Figure 3.7 (Plate 6)   Separation of catalyst from products using a short column of FRPSG
(Photograph by Ben Croxtall)

yse the reaction between acetoacetanoate and ethyl cyanoformate in good yield (Scheme 3.10).

Remarkably, passing the reaction mixture down a short column containing FRPSG and, eluting with dichloromethane, affords the analytically pure product (Figure 3.7). The catalyst (seen as the green band on top of the column) can then be quantitatively recovered by eluting with diethyl ether and, after washing, could be reused without loss in activity. This procedure was repeated five times.

3.8 CONCLUSIONS

Fluorous technology is a rapidly expanding area and has important and wide-ranging potential applications in both organic and inorganic chemistry. This thriving area of research, which only came into being in the early 1990s, now encompasses catalysis, solvents, nanoparticles and enzymes. A wide range of chemical reactions have been carried out in perfluorinated solvents. Some examples include hydroformylation [33], hydrogenation [34], Diels–Alder [35], allylic alkylations [36], oxidation [37], hydrosilylation [38], Heck, Suzuki and Stille reactions [39] and polymerizations [40]. Exploitation of this methodology is reliant on the adaptation of reagents and catalysts to the fluorous phase and the ability to separate them again, and developments are advancing rapidly. Although cost is a major issue with this fluorous methodology, it is expected that, as demand increases, costs will decrease. The use of a 'traditional' fluorous extraction on an industrial scale would need a huge volume of fluorous solvent and is therefore probably unlikely, but new techniques are becoming available which may make this unnecessary. It is likely that the fluorous techniques will find applications in areas which produce high value products, which would minimize cost issues.

REFERENCES

1. (a) Horváth I. T. and Rábai J. *Science* 1994, **266**, 72; (b) Horváth I. T. *Acc. Chem. Res.* 1998, **31**, 641.
2. Hildebrand J. H. and Cochran D. R. F. *J. Am. Chem. Soc.* 1949, **71**, 22.
3. Green S. W., Slinn D. S. L., Simpson R. N. F. and Woytek A. J. In *Organofluorine Chemistry – Principles and Commercial Application*, Banks R. E., Smart B. E. and Tatlow J. C. (eds), Plenum Press, New York, 1994, 89.
4. Freed B. K., Biesecker J. and Middleton W. J. *J. Fluorine Chem.* 1990, **48**, 63.
5. Riess J. G. and Le Blanc M. *Angew. Chem., Int. Ed. Engl.* 1978, **17**, 621.
6. Marcus Y. *Chem. Soc. Rev.* 1993, 409.
7. May G. *Chem. Br.*, 1997, 34.
8. Ravishankara A. R., Solomon S., Turnipseed A. A. and Warren R. F. *Science* 1993, **259**, 194.
9. Herrera V., de Rege P. J. F., Horváth I. T., Le Husebo T. and Hughes R. P. *Inorg. Chem. Commun.s* 1998, **1**, 197.
10. Alvey L. J., Meier R., Soos T., Bernatis P. and Gladysz J. A. *Eur. J. Inorg. Chem.* 2000, 1975.

11. Bhattacharyya P., Gudmunsen D., Hope E. G., Kemmitt R. D. W., Paige D. R. and Stuart A. M. *J. Chem. Soc. Perkin Trans. I* 1997, 3609.

12. Haar C. M., Huang J. and Nolan S. P. *Organometallics* 1998, **17**, 5018.

13. McLoughlin V. C. R. and Thrower J. *Tetrahedron* 1969, **25**, 5921.

14. (a) Bhattacharyya P., Gudmunsen D., Hope E. G., Kemmitt R. D. W., Paige D. R. and Stuart A. M. *J. Chem. Soc. Perkin Trans. I* 1997, 3609; (b) Mathivet T., Monflier E., Castanet Y., Mortreux A. and Couturier J.-L. *Tetrahedron Lett.*, 1999, **40**, 3885; (c) Chem W., Xu L., Hu Y., Banet Osuna A. M. and Xiao J. *Tetrahedron* 2002, **58**, 3889.

15. www.fluorous.com

16. (a) Kainz S., Koch D., Baumann W. and Leitner W. *Angew. Chem., Int. Ed. Engl.* 1997, **36**, 1628; (b) Kainz S., Luo Z., Curran D. P. and Leitner W. *Synthesis* 1998, 1425.

17. Zhang Q. S., Luo Z. Y. and Curran D. P. J. *J. Org. Chem.* 2000, **65**, 8866.

18. (a) Chem W., Xu L. and Xiao J. *Tetrahedron Lett.* 2001, **42**, 4275; (b) Darses S., Pucheault M. and Genet J.-P. *Eur. J. Org. Chem.* 2001, 1121.

19. (a) Soos T., Bennett B. L., Rutherford D., Barthel-Rosa L. P. and Gladysz J. A. *Organometallics* 2001, **20**, 3079.

20. Rocaboy C., Hampel F., and Gladysz J. A. *J. Org. Chem.* 2002, **67**, 6863.

21. (a) Sinou D., Pozzi G., Hope E. G. and Stuart A. M. *Tetrahedron Lett.* 1999, **40**, 849; (b) Sinou D., Maillard D. and Pozzi G. *Eur. J. Org. Chem.* 2002, 269; (c) Cavazzini M., Pozzi G., Quici S., Maillard D. and Sinou D. *Chem. Commun.*, 2001, 1220.

22. de Wolf E., Richter B., Deelman B.-J. and van Koten G. *J. Org. Chem.* 2000, **65**, 5424.

23. (a) Fawcett J., Hope E. G., Kemmitt R. D. W., Paige D. R., Russell D. R. and Stuart A. M. *J. Chem. Soc. Dalton Trans.* 1998, 3751; (b) Hope E. G., Kemmitt R. D. W., Paige D. R., Stuart A. M. and Wood D. R. W. *Polyhedron* 1999, **18**, 2913; (c) Croxtall B., Fawcett J., Hope E. G. and Stuart A. M. *J. Chem. Soc. Dalton Trans.* 2002, 491.

24. (a) Adams D. J., Gudmunsen D., Fawcett J., Hope E. G. and Stuart A. M. *Tetrahedron* 2002, **58**, 3827; (b) Adams D. J., Fawcett J. and Hope E. G., unpublished results.

25. (a) www.organik.uni-erlangen.de/gladysz/research/gifs/fluorous/partition.html; (b) Chen W., Xu L., Hu Y., Bant-Osuna A. M. and Xiao J. *Tetrahedron* 2002, **58**, 3889.

26. (a) Horváth I. T. and Rábai J. *Science* 1994, **266**, 72; (b) Horváth I. T., Kiss G., Cook R. A., Bond J. E., Stevens P. A. and Rábai J. *J. Am. Chem. Soc.* 1998, **120**, 3133.

27. (a) Foster D. F., Gudmunsen D., Adams D. J., Stuart A. M., Hope E. G. and Cole-Hamilton D. J. *Chem. Commun* 2002, 722; (b) Foster D. F., Gudmunsen D., Adams D. J., Stuart A. M., Hope E. G., Cole-Hamilton D. J., Schwarz G. P. and Pogorzelec P. *Tetrahedron* 2002, **58**, 3901.

28. Wende M., Meier R. and Gladysz J. A. *J. Am. Chem. Soc.* 2001, **123**, 11490.

29. Curran D. P. *Synlett.* 2001, 1488.

30. Curran D. P. and Lee Z. *Green Chem.* 2001, **3**, G3.

31. Luo Z., Zhang Q., Oderaotoshi Y. and Curran D. P. *Science* 2001, **291**, 1766.

32. Croxtall B., Hope E. G. and Stuart A. M. *Abstr. RSC Ann. Conf.*, 2001, FBC7.

33. (a) Horváth I. T. and Rábai J. *Science*, 1994, **266**, 72; (b) Horváth I. T., Kiss G., Cook R. A., Bond J. E., Stevens P. A. and Rábai J. *J. Am. Chem. Soc.* 1998, **120**, 3133; (c) Foster D. F., Gudmunsen D., Adams D. J., Stuart A. M., Hope E. G. and Cole-Hamilton D. J. *Chem. Commun.* 2002, 722; (d) Foster D. F., Gudmunsen D., Adams D. J., Stuart A. M., Hope E. G., Cole-Hamilton D. J., Schwarz G. P. and Pogorzelec P. *Tetrahedron* 2002, **58**, 3901.

34. Rutherford D., Juliette J. J. J., Rocaboy C., Horváth I. T. and Gladysz J. A. *Catal. Today* 1998, **42**, 381.

35. Myers K. E. and Kumar K. *J. Am. Chem. Soc.* 2000, **122**, 12025.
36. (a) Cavazzini M., Pozzi G., Quici S., Maillard D. and Sinou D. *Chem. Commun.*, 2001, 1220; (b) Maillard D., Bayardon J., Kurichiparambil J. D., Nguefack-Fournier C. and Sinou D. *Tetrahedron. Asymmetry* 2002, **13**, 1449.
37. (a) Pozzi G., Cavazzini M., Cinato F., Montanari F. and Quici S. *Eur. J. Org. Chem.* 1999, 1947; (b) Crich D. and Neelamkavil S. *J. Am. Chem. Soc.* 2001, **123**, 7449; (c) Betzemeier B., Cavazzini M., Quici S. and Knochel P. *Tetrahedran Lett.* 2000, **41**, 4343.
38. de Wolf E., Speets E. A., Deelman B.-J. and van Koten G. *Organometallics*, 2001, **20**, 3686.
39. (a) Moreno-Manas M., Pleixats R. and Villarroya S. *Organometallics* 2001, **20**, 4524; (b) Schneider S. and Bannwarth W. *Angew. Chem., Int. Ed. Engl.* 2000, **39**, 4142; (c) Moineau J., Pozzi G., Quici S. and Sinou D. *Tetrahedran Lett.* 1999, **40**, 7683; (d) Schneider S. and Bannwarth W. *Helv. Chim. Acta* 2001, **84**, 735.
40. Haddleton D. M., Jackson S. G. and Bon S. A. F. *J. Am. Chem. Soc.* 2000, **122**, 1542.

4 Ionic Liquids

The first example of biphasic catalysis was actually described for an *ionic liquid* system. In 1972, one year before Manassen proposed aqueous–organic biphasic catalysis [1], Parshall reported that the hydrogenation and alkoxycarbonylation of alkenes could be catalysed by $PtCl_2$ when dissolved in tetraalkylammonium chloride/tin dichloride at temperatures of less than $100\,^\circ C$ [2]. It was even noted that the product could be separated by decantation or distillation. Since this nascent study, synthetic chemistry in ionic liquids has developed at an incredible rate. In this chapter, we explore the different types of ionic liquids available and assess the factors that give rise to their low melting points. This is followed by an evaluation of synthetic methods used to prepare ionic liquids and the problems associated with these methods. The physical properties of ionic liquids are then described and a summary of the properties of ionic liquids that are attractive to clean synthesis is then given. The techniques that have been developed to improve catalyst solubility in ionic liquids to prevent leaching into the organic phase are also covered.

4.1 INTRODUCTION

Ionic liquids have been known since 1914 when ethylammonium nitrate, $[EtNH_3][NO_3]$, with a melting point of $12\,^\circ C$ was discovered by Walden [3]. For the next forty years there was no activity in this field, until other low melting ionic compounds were discovered by chemists looking for an alternative method to electroplate aluminium. On mixing and heating various alkylpyridinium chlorides with aluminium chloride, the white powders reacted together to form a colourless liquid [4]. However, ionic liquids remained something of a curiosity until only recently when they were rediscovered as alternatives to common organic solvents for synthetic applications.

There are numerous ways in which an ionic liquid can be defined, and perhaps the most widely accepted definition is:

A material that is composed of ions, and has a melting point below $100\,^\circ C$.

Such a definition would suggest that hundreds or even thousands of compounds could be classed as ionic liquids, but the number is in fact relatively small.

Chemistry in Alternative Reaction Media   D. Adams, P. Dyson and S. Tavener
© 2004 John Wiley & Sons, Ltd   ISBNs: 0-471-49848-3 (Cloth); 0-471-49849-1 (Paper)

However, over the last few years the number of known ionic liquids has risen rapidly and potentially millions of different ionic liquids could exist, as shall become evident later in this chapter.

Ionic liquids are also sometimes referred to as *molten salts, nonaqueous ionic liquids* (NAILs) or *room temperature ionic liquids,* and all of these names are entirely valid. The term molten salt is now used less frequently, and generally

---

What Controls the Melting Point of an Ionic Compound?

When one thinks of an ionic compound, i.e. a salt, one usually thinks of a crystalline solid with a high melting point. For example, sodium chloride has a high melting point (806 °C) because its ions are held together by strong electrostatic attractions leading to a high lattice energy (see Figure 4.1). The strength of electrostatic attraction is determined by the amount of charge on the ions and how far away from each other they are. The further that they are apart, then the weaker the interaction. So how do you separate two things that are attracted to each other strongly? It is possible to make ions that have their charge on a central atom and then have large groups bonded to them so that nothing else can get close. This greatly increases the distance between the charges (e.g. [Bu$_4$N][Br] has a melting point of just over 100 °C). The same applies to the anions, and replacing the anion with a bulkier one can have an even greater effect on melting point. In addition, by delocalizing the charge over many atoms instead of having one particular point in the ion with a strong attraction, there are lots of places with just a little attraction and in different directions. This is very effective at weakening the force between the two ions. Lastly, cations with low symmetry may be used, so that it is even more difficult for the charges to come together, this also helps to reduce melting points.

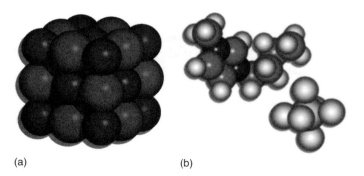

(a)   (b)

Figure 4.1 (Plate 7)   Sodium chloride (a) can easily form a regular lattice through packing of the spherical anions and cations. It is much harder for [bmim][PF$_6$] (b) to do so because the charges are dispersed across the large ions, and the cation has low symmetry refers to salts with melting points greater than 100 °C. NAILs was originally used to differentiate synthetic ionic liquids from water, since liquid water is partially composed of ions. The last term, room temperature ionic liquids, is sometimes used to differentiate those that are truly low melting from molten salts with high melting points. Although the definition given above includes any ionic material with a melting point below 100 °C, it is really those that are liquid at, or below, room temperature that are most useful as alternative solvents for clean synthesis.

4.1.1   The Cations and Anions

Ionic liquids may be viewed as three-dimensional networks of cations and anions linked together by interactions such as H-bonds, dispersive and electrostatic forces. The key feature of the cation seems to be that it should be of low symmetry and have a poorly localized positive charge, impeding the formation of a regular crystal lattice and lowering the melting point of the subsequent salt. There are now many ionic liquids based on organic cations that can be coupled to a wide range of anions to provide solvents with specific chemical and physical properties. The most commonly used cations that are well suited to giving low melting salts are illustrated in Figure 4.2 [5]. These cations are related in that they are of quite low symmetry and the charge on many of them is delocalized over several atoms.

Many anions can be used in combination with the cations shown in Figure 4.2 to form low melting liquids, including metal halide anions, BF_4^-, PF_6^-, AsF_6^-, SbF_6^-, NO_3^-, $CH_3CO_2^-$, $CF_3SO_3^-$, $(CF_3SO_2)_2N^-$, $(CF_3SO_2)_3C^-$ and $CF_3CO_2^-$ to name but a few. However, the most important component in an ionic liquid is the cation. Whereas certain cations may be combined with a wide range of anions to give low melting salts there is not, as yet, a universal 'liquifying' anion that can be combined with different inorganic and organic cations (excluding those shown in Figure 4.2) to form low melting salts.

4.1.2   Synthesis of Ionic Liquids

The synthetic routes used to prepare ionic liquids vary depending upon the ionic liquid being made. Ionic liquids with metal halide anions are, at least in principle, very simple to prepare. Scheme 4.1 illustrates the synthesis of imidazolium-based ionic liquids with a chloroaluminate anion, commencing with methylimidazole [6].

Although the synthesis appears to be very straightforward, it is actually very difficult to obtain a colourless liquid, as trace amounts of impurities lead to discolouration. The first step of the reaction does not go to completion and the

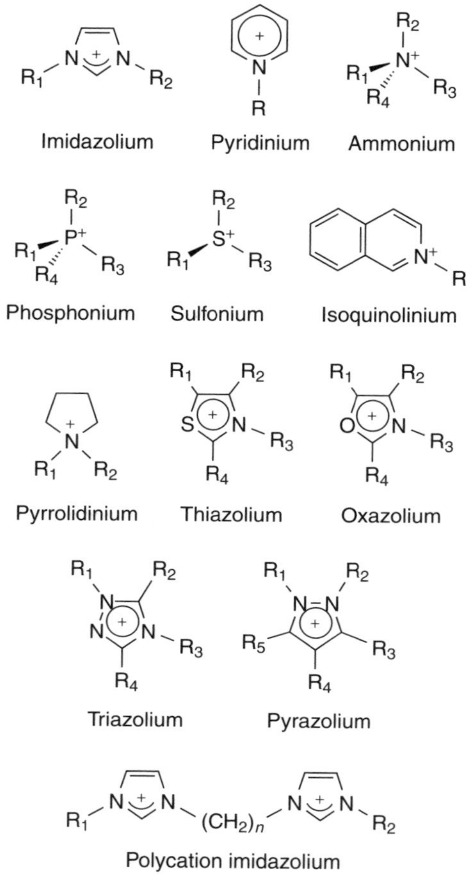

Figure 4.2 Some cations that give rise to low melting salts; R is usually an alkyl group although various functional groups can also be used

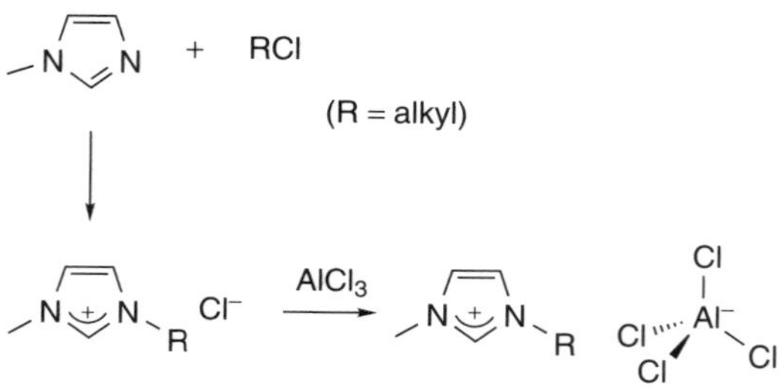

Scheme 4.1 The synthesis of imidazolium–chloroaluminate ionic liquids (where R is an alkyl group)

product must be recrystallized several times to remove all the starting material. The imidazolium salt and the aluminium trichloride used in the second step must both be completely dry. The aluminium trichloride must be of high purity, which may be achieved by sublimation. In general, the pure and dry white solid starting materials are mixed together under an inert atmosphere and they collapse to form a colourless liquid. The imidazolium chloride and the aluminium trichloride need not be mixed in a 1 : 1 molar ratio in order to form a liquid and the consequence of mixing different molar ratios has a marked influence on the nature of the anion present and the subsequent properties of the ionic liquid [7]. Mixing equimolar amounts of the chloride salt and $AlCl_3$ gives a mole fraction of 0.5 and the $AlCl_4^-$ anion is essentially the only species present. If the mole fraction of $AlCl_3$ employed is greater than 0.5, then multinuclear species such as $Al_2Cl_7^-$ and $Al_3Cl_{10}^-$ are formed and the ionic liquid is referred to as (Lewis) acidic. When the mole fraction is less than 0.5, the ionic liquid is basic. For this reason, chloroaluminate ionic liquids are often written as [cation]Cl-$AlCl_3$, rather than [cation]$AlCl_4$ and the mole fraction is quoted at the same time. The ability to prepare the ionic liquid in neutral, acidic or basic form may be exploited in synthesis and catalysis.

The synthesis of ionic liquids with BF_4^- and PF_6^- as cations has been the subject of much research since they are the most widely used in catalysis. However, it is difficult to make these ionic liquids in a pure form. The original route used to prepare ionic liquids with these anions consists of a metathesis (anion–cation exchange) reaction in which the imidazolium chloride is reacted with the sodium salt of the anion in a suitable solvent [8]. The reaction is illustrated in Scheme 4.2 for the tetrafluoroborate salt.

Once the two salts are mixed in solution (acetone is a common solvent for this), the sodium chloride precipitates and is removed by filtration. The solvent is then removed under reduced pressure and, since salts have no vapour pressure, the ionic liquid remains in the flask. The problem with this reaction is that it is almost impossible to remove the last traces of chloride ions. The chloride not only influences the physical properties of the liquid such as melting point and viscosity, but is also a good nucleophile and can deactivate catalysts and affect reproducibility. A great deal of effort has been directed towards removal of the chloride contamination, including washes and chromatography, but none have proved to be completely effective [9]. This has led to the development of some alternative synthetic routes. Simply exchanging Na[BF_4]

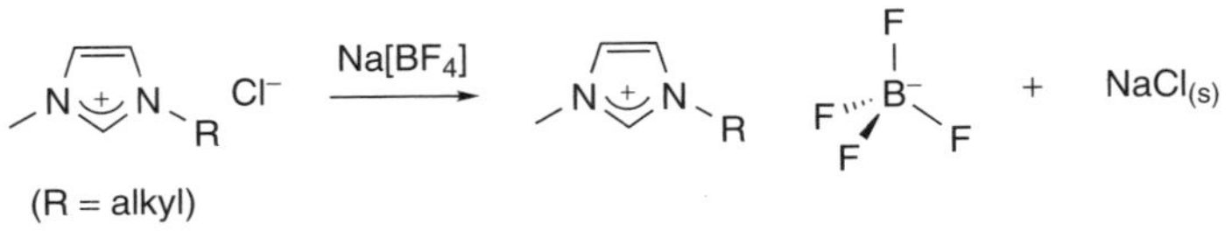

(R = alkyl)

Scheme 4.2 The preparation of imidazolium tetrafluoroborate ionic liquids (where R is an alkyl group)

(R = alkyl)

Scheme 4.3 One step synthesis of imidazolium tetrafluoroborate ionic liquids from alkylimidazoles (where R is an alkyl group)

(R = alkyl)

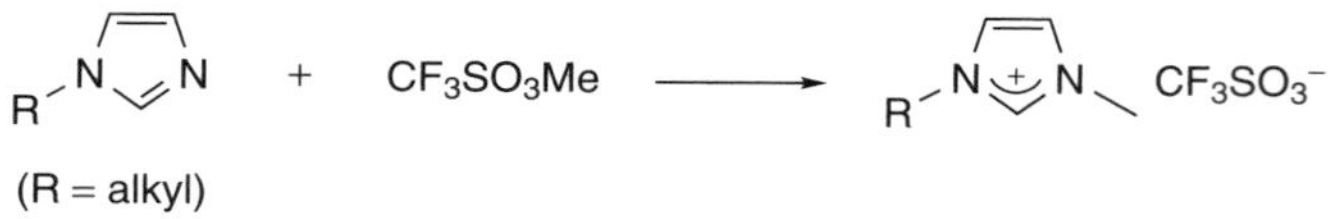

Scheme 4.4 Synthesis of imidazolium triflate ionic liquids from alkylimidazoles (where R is an alkyl group)

for the silver salt, Ag[BF$_4$], increases the yield of the precipitate and gives a purer product, because the lattice energy of AgCl is high compared to that of NaCl [10]. However, this route is very expensive and some chloride contamination still remains. Use of fluoroboric acid, HBF$_4$, results in the formation of HCl, which can be removed as a gas under reduced pressure rather than by precipitation [11]. However, complete removal of the HCl is not without problems as it is highly corrosive.

Perhaps the most effective synthesis, albeit an expensive one, involves the direct methylation of alkylimidazoles using trimethyloxonium tetrafluoroborate with concomitant formation of the tetrafluoroborate anion as shown in Scheme 4.3 [12]. This is a single step reaction in which no chloride is present at any stage of the synthesis. The only by-product from this reaction is dimethyl ether which is a volatile and inert gas and it is easily removed.

Many ionic liquids with other anions are produced in related one-pot methylation reactions that do not give rise to by-products [13]. For example, methyl triflate, CF$_3$SO$_3$Me, can be used to methylate alkylimidazoles to give 1,3-alkylmethylimidazolium triflate ionic liquids as shown in Scheme 4.4.

4.2 PHYSICAL PROPERTIES OF IONIC LIQUIDS

The physical properties of ionic liquids have been extensively studied and some trends are beginning to emerge. In particular, ionic liquids based on 1,3-dialkylimidazolium cations have been investigated in detail, partly due the their wide use as solvents to conduct synthesis and catalysis. The attraction of the imidazolium cation in synthetic applications is because the two substituent groups can be varied to modify the properties of the solvent. For example, Table 4.1

Table 4.1 Effect of the R group on the melting point of some imidazolium-based ionic liquids

Imidazolium tetrafluoroborate salt		Mp (°C)
	[mmim][BF$_4$]	103
	[emim][BF$_4$]	6
	[bmim][BF$_4$]	−81
	[hmim][BF$_4$]	−82
	[omim][BF$_4$]	−78
	[dmim][BF$_4$]	−4

Table 4.2 Effect of the anion on the melting point of the 1-butyl-3-methylimidazolium, [bmim]$^+$, cation

Anion	Mp (°C)
Cl$^-$	65
I$^-$	−72
BF$_4{}^-$	−81
PF$_6{}^-$	−61
AlCl$_4{}^-$	65
$(n = 0.66)^a$	
CF$_3$SO$_3{}^-$	16
(CF$_3$CO$_2{}^-$)	−50
(CF$_3$SO$_2$)$_2$N$^-$	−4

^{a} n is the mole fraction of [bmim]Cl:AlCl$_3$

shows how the alkyl group attached to the imidazolium cation influences the melting point of the resulting ionic liquid [14].

The cation with the highest symmetry, [mmim], has a higher melting point than the other less symmetric cations. In addition, as the chain length increases beyond an ethyl group there is little change in melting point until it reaches 10 carbon atoms and the melting point starts to increase again, presumably caused by attractions between the alkyl chains. The nature of the anion also has a major influence on both the physical and chemical properties of the ionic liquid (see

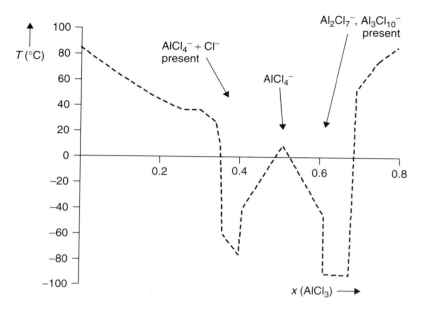

Figure 4.3   Phase diagram for [emim]Cl-AlCl$_3$

below). The influence of the anion on the melting point of a series of [bmim]$^+$ ionic liquids is shown in Table 4.2. The effect of the anion on the melting point is not easy to predict although in general the melting points provided by smaller anions are higher than those obtained with larger more bulky anions.

The properties of the chloroaluminate ionic liquids can be modified by varying the mole fraction of the [cation]Cl and AlCl$_3$. When the mole fraction of [cation]Cl:AlCl$_3$ is 0.5 the AlCl$_4$$^-$ anion is essentially the only species present and the ionic liquid is neutral. If the mole fraction of AlCl$_3$ employed in greater than 0.5, then the ionic liquid is Lewis acidic and when the mole fraction is less than 0.5 the ionic liquid is Lewis basic. The melting point of the liquid depends on the composition and the phase diagram shown in Figure 4.3 illustrates how the melting point varies with molar ratio.

The polarity of some 1-alkyl-3-methylimidazolium ionic liquids has been determined using the solvatochromic dyes Nile Red and Reichardt's dye [16, 17]. Measurements with Nile Red do not give absolute values of polarity but provide a useful scale to estimate the relative polarity of the ionic liquids. Similar measurements have been made using a number of other solvatochromic dyes (dansylamide, pyrene, pyrenecarboxyaldehyde, and bromonapthalene) for [BMIM][PF$_6$], and gave results consistent with those obtained with Nile Red. Values for E_T^N obtained for ionic liquids generally fall between the values of 0.6 and 1.0, as shown in Table 4.3.

Zwitterionic Liquids

A number of zwitterionic liquids in which an imidazolium cation is directly bonded to the anion are also known [15]. The melting point for these zwitterionic liquids is very low. Although these zwitterionic liquids have high ion density, the ions cannot migrate along potential gradients and the ionic conductivity of pure zwitterionic liquid is quite low. However, when mixed with an equimolar amount of the salt lithium bis(trifluoromethanesulfonyl)imide (LiTFSI), the zwitterionic liquids display a high ionic conductivity. Therefore the zwitterionic liquids act as an excellent ion conductive matrix, in which only added ions can migrate. Zwitterionic liquids with vinyl groups attached to the imidazolium rings can be polymerized. The ionic conductivity of pure zwitterionic polymers is again low, but when mixed with LiTFSI these polymers exhibit a relatively high ionic conductivity despite having rubber-like properties. Both zwitterionic-type salts and their polymers have unique ion conductive characteristics that may find uses in electrochemistry and as novel conducting materials.

Using Nile Red, the polarity of several ionic liquids has been found to be comparable to that of lower alcohols. Figure 4.4 summarizes the polarity data for some ionic liquids with comparisons to common organic solvents (the structure of Nile Red is shown in Chapter 1). Polar organic solvents like dichloromethane and diethyl ether are miscible with ionic liquids, whereas solvents of low polarity show partial miscibility and nonpolar solvents are essentially immiscible.

Comparative polarity studies have also been performed by absorption of ionic liquids on to gas chromatography columns followed by the elution of various compounds and lead to the conclusion that the polarity of [bmim][BF$_4$] is similar to that of lower alcohols [18].

Miscibility is an important consideration when selecting solvents for use in biphasic systems. Table 4.4 shows the miscibility of three ionic liquids with water and some organic solvents. [bmim][PF$_6$] was found to be miscible with organic solvents whose dielectric constant is higher than 7, but was not soluble in less polar solvents or in water. Basic [bmim][AlCl$_4$] was found to react with protic solvents, and the acidic form also reacted with acetone, tetrahydrofuran and toluene.

Table 4.3 E_T^N values for a range of ionic liquids [17]

Solvent	E_T^N
—N⁺N—Bu [BF₄]⁻	0.58
[bmim][O₂CCF₃]	0.62
[bmim][N(SO₂CF₃)₂]	0.65
—N⁺N—C₃H₇ ⁻N(SO₂CF₃)₂	0.66
—N⁺N—C₁₀H₂₁ ⁻N(SO₂CF₃)₂	0.66
[bmim][OSO₂CF₃]	0.66
[bmim][BF₄]	0.67
[bmim][PF₆]	0.67
—N⁺N—CH₂Ph ⁻N(SO₂CF₃)₂	0.67
[bmim][ClO₄]	0.68
—N⁺N—(CH₂)₂OCH₃ ⁻N(SO₂CF₃)₂	0.72
—N⁺N—(CH₂)₂OH ⁻N(SO₂CF₃)₂	0.95
[Et₃NH]NO₃	0.95

The proton in the 2-position of imidazolium-based ionic liquid is quite acidic, as evidenced both spectroscopically and from its reactivity (direct pH measurements in ionic liquids using a pH meter do not give meaningful data). Such acidity would suggest that the cation can form hydrogen bonds with the anions as well as with compounds dissolved in the ionic liquid. A number of X-ray diffraction studies have been conducted to probe the solid-state structure of various 1,3-dialkylimidazolium salts revealing the presence of extended H-bonded networks although the shortest interaction does not always involve the proton in the 2-position [19]. Evidence for H-bonding in ionic liquids in the solution-state has also been provided by NMR studies [20]. The presence of H-bonding is clearly important in determining the melting point and viscosity of an ionic liquid, although a number of other factors are also important. What is unusual, based on the acidity of the 2-proton, is that when it is replaced by a methyl group the melting point of the ionic liquid increases rather than decreases, as illustrated in Table 4.5.

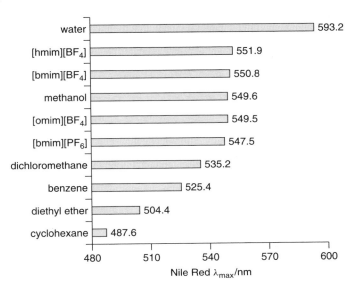

Figure 4.4 The solvatochromic shifts of the dye Nile Red in a number of ionic liquids and organic solvents

Table 4.4 Miscibility of organic solvents with ionic liquids

Organic solvent	ε_r	[bmim][PF$_6$]	[bmim][AlCl$_4$] Basic	[bmim][AlCl$_4$] Acidic
Water	78.3	Immiscible	Reacting	Reacting
Propylene carbonate	64.4	Miscible	Miscible	Miscible
Methanol	32.7	Miscible	Reacting	Reacting
Acetonitrile	35.9	Miscible	Miscible	Miscible
Acetone	20.6	Miscible	Miscible	Reacting
Methylene chloride	8.9	Miscible	Miscible	Miscible
Tetrahydrofuran	7.8	Miscible	Miscible	Reacting
Trichloroethylene	3.4	Immiscible	Immiscible	Immiscible
Carbon disulfide	2.6	Immiscible	Immiscible	Immiscible
Toluene	2.4	Immiscible	Miscible	Reacting
Hexane	1.9	Immiscible	Immiscible	Immiscible

Source: Koel M. *Proc. Estonian Acad. Sci. Chem.* 2000, **49**, 145.

So far, we have focused on the melting points and polarities of ionic liquids. Like conventional solvents, other properties such as viscosity and density are also very important when selecting a solvent for synthetic applications. Whilst this type of data is well known for other solvents, relatively little has been reported for ionic liquids. Table 4.6 lists available melting points, thermal stability, density, viscosity and conductivity data for the better studied ionic liquids.

Table 4.5 A comparison of the
melting points of
1-ethyl-3-methylimidazolium and
1-ethyl-2,3-dimethylimidazolium salts

Ionic liquid or salt	Mp ($^\circ$C)
Cl^-	81
Cl^-	187
$CF_3SO_3^-$	-9
$CF_3SO_3^-$	109

Table 4.6 Some physical properties of imidazolium-based ionic liquids

Cation	Anion	Mp ($^\circ$C)	Thermal stability (max. liquid temp.)($^\circ$C)	Density (g cm^{-3}) [Temp.]	Viscosity (cP) [Temp.]	Conductivity (ohm^{-1} cm^{-1}) [Temp.]
H_3C— —CH_3	$(CF_3SO_2)_2N^-$	22	–	1.559 [22]	444 [20]	0.84 [20]
C_2H_5— —CH_3	BF_4^-	6	412	1.24 [22]	37.7 [22]	1.4 [25]
C_2H_5— —CH_3	$CF_3SO_3^-$	-9	~440	1.38 [25]	42.7 [25]	0.92 [25]
C_2H_5— —CH_3	$CF_3CO_2^-$	-14	~150	1.285 [22]	35 [20]	0.96 [20]
C_2H_5— —CH_3	$(CF_3SO_2)_2N^-$	-3	455	1.520 [22]	34 [22]	0.88 [25]
C_2H_5— —C_2H_5	$CF_3SO_3^-$	23	–	1.330 [22]	53 [20]	0.75 [20]

Table 4.6 (continued)

Cation	Anion					
C_2H_5–N⁺N–C_2H_5 (H)	$CF_3CO_2^-$	−50	–	1.250 [22]	43 [20]	0.74 [20]
C_2H_5–N⁺N–C_2H_5 (H)	$(CF_3SO_2)_2N^-$	14	–	1.452 [21]	35 [20]	0.85 [20]
C_4H_9–N⁺N–CH_3 (H)	BF_4^-	−81	403	1.12 [25]	219 [25]	0.173 [25.5]
C_4H_9–N⁺N–CH_3 (H)	PF_6^-	−61	349	1.36 [25]	450 [25]	0.146 [25.5]
C_4H_9–N⁺N–CH_3 (H)	$(CF_3SO_2)_2N^-$	−4	439	1.429 [19]	52 [20]	0.39 [20]
C_4H_9–N⁺N–C_2H_5 (H)	$CF_3(CF_2)_2SO_3^-$	21	–	1.427 [18]	323 [20]	0.053 [20]
C_4H_9–N⁺N–C_2H_5 (H)	$CF_3CO_2^-$	−50	–	1.428 [20]	89 [20]	0.25 [20]
C_6H_{11}–N⁺N–CH_3 (H)	PF_6^-	−61	417	1.29 [25]	585 [25]	–
C_2H_5–N⁺N–CH_3 (CH_3)	$CF_3SO_3^-$	6	–	1.334 [20]	51 [20]	0.64 [20]
C_2H_5–N⁺N–CH_3 (CH_3)	$(CF_3SO_2)_2N^-$	−3	–	1.470 [22]	37 [20]	0.66 [20]

The solubility of organic compounds in ionic liquids can be estimated from the polarity data given in Table 4.3 and Figure 4.4. In general, solids are of limited solubility in ionic liquids unless they are salts themselves and then solubility is generally high. The solubility of gases is also of importance as many catalysed reactions that are well suited to biphasic processes involve gaseous substrates such as hydrogen, carbon monoxide and carbon dioxide. The Henry's constant for some gases in ionic liquids is given in Table 4.7 [21].

Table 4.7 Solubility of gases in ionic liquids at 25 °C

Gas–ionic liquid	Henry's constant
CO_2 – [bmim][PF_6]	53.4
C_2H_4 – [bmim][PF_6]	173
C_2H_6 – [bmim][PF_6]	355
CH_4 – [bmim][PF_6]	1690
O_2 – [bmim][PF_6]	8000
Argon – [bmim][PF_6]	8000
H_2 – [bmim][PF_6]	14 925
H_2 – [bmim][CF_3SO_3]	6896
H_2 – [bmim][BF_4]	6923
H_2 – [hmim][BF_4]	11 627
H_2 – [omim][BF_4]	16 393

The lower the value for Henry's constant, the greater the solubility of the gas in the solvent. It is clear from the data in Table 4.7 that CO_2 is very soluble, whereas the solubility of hydrogen in all the ionic liquids is very low. In fact, the Henry's constant for all the gases except for hydrogen were determined using a gravimetric microbalance at 13 atm, but at this pressure hydrogen was below the detection limit. The Henry's constants for hydrogen were determined at 100 atm using high pressure ^{1}H NMR spectroscopy. Despite the low solubility of certain gases the rate of mass transport is very high in ionic liquids and gas solubility does not appear to be a limiting factor in catalysed reactions involving these gases.

4.3 BENEFITS AND PROBLEMS ASSOCIATED WITH USING IONIC LIQUIDS IN SYNTHESIS

Ionic liquids have a number of properties that make them suitable media for conducting chemical synthesis and catalysis [22]:

1 They dissolve many metal catalysts, polar organic compounds and gases and can even support biocatalysts.
2 They have favourable thermal stabilities and operate over large ranges. Most melt below room temperature and only start to decompose above 300 or 400 °C which gives a temperature range three to four times that of water in which to conduct synthesis.
3 They may be designed to be immiscible with many organic solvents and water, and both the cations and anions can be modified to give specific solubility properties as well as other physical or chemical properties.
4 Ionic liquids have polarities comparable to lower alcohols, but, unlike alcohols and other polar solvents, they are essentially non-coordinating (occasionally

weakly coordinating, depending on the anion). The fact that they do not coordinate to metal centres is advantageous when the ionic liquid is used to immobilize sensitive transition metal catalysts.

5   They have no vapour pressure and therefore do not evaporate. This means they do not escape into the environment like volatile organic solvents and it also allows easy removal of VOCs (i.e. reaction products) from the ionic liquid under vacuum or by distillation.

6   While very little toxicity data are available it would appear that many ionic liquids are nontoxic.

7   Ionic liquids in which the anion is composed of an active transition metal catalyst providing essentially a 'liquid catalyst' may be prepared, although it is too early to comment whether these will offer any advantages over catalysts dissolved in ionic liquids [23].

Essentially, there is no limit to the number of different ionic liquids that can be engineered with specific properties for chemical applications. However, a number of problems still need to be overcome before their use becomes widespread. The current problems associated with ionic liquids include:

1   Many are difficult to prepare in a pure form, and the current methods that provide pure ionic liquids are generally very expensive. Scale-up could be a problem in certain cases.

2   The viscosity of ionic liquids is often quite high. In addition, impurities can have a marked influence and may increase the viscosity of the ionic liquid. In the worse case scenario the addition of a catalyst and substrate to an ionic liquid can increase the viscosity to such an extent that it becomes gel-like and therefore difficult to process.

3   Some ionic liquids (e.g. chloroaluminates) are highly sensitive to oxygen and water, which means they can only be used in an inert environment and that all substrates must be dried and degassed before use.

4   Catalysts immobilized in ionic liquids are sometimes leached into the product phase. It may therefore be necessary to design new catalysts for use in ionic liquids.

Despite these problems, ionic liquids are currently attracting considerable attention as alternatives to volatile organic solvents in many different reactions including oligomerization and polymerization, hydrogenation, hydroformylation and oxidation, C–C coupling and metathesis. In particular, ionic liquids containing BF_4^- or PF_6^- anions have been very widely used and several general properties have emerged:

1   They form separate phases with many organic materials and can therefore be used in biphasic catalysis.

2   They are non-nucleophilic and present an inert environment that often increases the lifetime of the catalyst.

3 The rate of diffusion of gases is very high compared to many conventional
   solvents and this leads to increased reaction rates in catalysed reactions
   involving gaseous substrates such as hydrogenation, hydroformylation and
   oxidation.

4.4 CATALYST DESIGN

Many catalysts originally developed to operate in organic solvents have since
been used in ionic liquids without any modification. In the case of neutral catalysts, leaching from the ionic liquid phase into the product phase can often be a
problem, whereas charged catalysts remain immobilized during product extraction
although some loss can still occur depending on the solvent used as the second
phase. With this in mind, a number of ligands have been made that are salts and
when attached to a neutral metal will render the resulting catalyst preferentially
soluble in the ionic liquid phase.

A class of aryl-phosphine ligands with sulfonium anions attached to aryl-
rings have been designed for use in aqueous–organic biphasic catalysis [24],
(see Chapter 5). For applications in water they are usually stabilized as Na$^+$ or
K$^+$ salts. Since these ligands are salts they may be used in ionic liquid–organic
biphasic catalysis, but usually with the Group 1 metal cation replaced with the
cation of the actual ionic liquid. This provides excellent solubility and retention
in the ionic liquid phase during product extraction. Other charged groups have
been incorporated into ligands and Figure 4.5 shows several which have been
developed for use in ionic liquid biphasic catalysis.

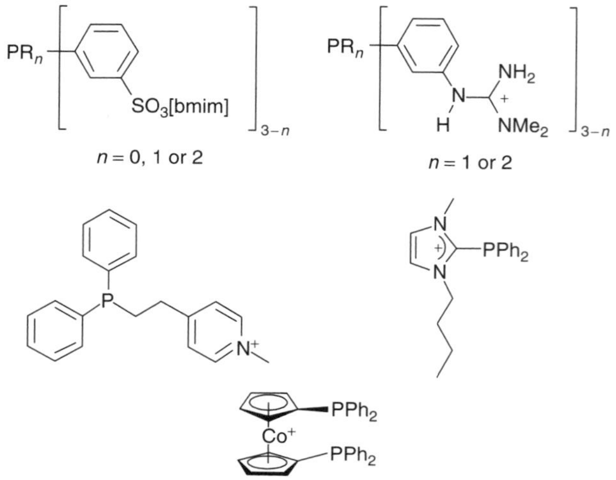

Figure 4.5 Ligands used to improve immobilization of metal catalysts in ionic liquids

The range of ligands developed for ionic liquid catalysis is much smaller than that for other immobilization solvents such as water and fluorous phases as 'off the shelf' ligands and catalysts can often be used in ionic liquids. For example, a number of catalysts that were developed to operate in organic solvents under homogeneous conditions are salts themselves and do not need to be modified for use in ionic liquids [25].

Catalysts other than homogeneous (molecular) compounds such as nanoparticles have been used in ionic liquids. For example, iridium nanoparticles prepared from the reduction of [IrCl(cod)$_2$] (cod = cyclooctadiene) with H$_2$ in [bmim][PF$_6$] catalyses the hydrogenation of a number of alkenes under biphasic conditions [27]. The catalytic activity of these nanoparticles is significantly more effective than many molecular transition metal catalysts operating under similar conditions.

Biocatalysts also operate in ionic liquids [28]. The ones that have been most widely investigated are the lipase family of enzymes. For example, *Candida Antarctica* lipase B immobilized in [bmim][BF$_4$] or [bmim][PF$_6$] under anhydrous conditions is able to catalyse transesterifications at rates comparable to those observed in other solvents. Certain lipase mediated enantioselective acylations have even resulted in considerable improvements in enantiomeric excesses

Chiral Ionic Liquids

Chiral ligands attached to metals centres have revolutionized the synthesis of chiral products. In 2001, the Nobel Prize for Chemistry was awarded to the founders of this field. There has been much speculation as to whether a nonchiral catalyst operating in a chiral solvent would also give products with enantiomeric excesses. With organic solvents, the results have been disappointing. But the potential for long range structure in an ionic liquid is greater and a number of chiral ionic liquids have been prepared [26], for example:

R$_1$ = Me, R$_2$ = Et,
R$_1$ = Pentyl, R$_2$ = Et

Thus far, these chiral ionic liquids do not appear to exert an influence, possibly because the structuring of the solvents around the reactions' transition state is not rigid enough. These investigations are still in progress.

compared to other media. As yet, while this nascent field has shown some startling results, little is understood regarding the effects induced by the ionic liquids.

4.5 CONCLUSIONS

Ionic liquids have many properties that lend themselves to clean chemical synthesis, perhaps most notably, that they are nonvolatile and are therefore not lost to the atmosphere. While many of their physical properties have yet to be determined a number of trends are emerging that allows the design of ionic liquids with specific properties, which will ultimately allow them to be fine tuned to specific processes. They have also been extensively studied as alternative solvents for organic chemistry and biphasic catalysis, and numerous examples of this aspect are given in Chapters 7–11. In addition to ionic liquid–organic and ionic liquid–aqueous biphasic catalysis a number of multiphasic combinations have been developed which show promise. One particularly promising combination is the synthesis of organic compounds under homogeneous conditions in an ionic liquid followed by extraction of the products with supercritical CO_2 [29]. In such a process all volatile organic solvents are eliminated.

REFERENCES

1. Manassen J. In *Catalysis: Progress in Research*, Bassolo F. and Burwell R. L. (eds), Plenum Press, London, 1973, p. 183.
2. Parshall G. W. *J. Am. Chem. Soc.*, 1972, **94**, 8716.
3. Sugden S. and Wilkins H. *J. Chem. Soc.* 1929, 1291, and references cited therein.
4. Hurley F. H. and Wier Jr. T. P. *J. Electrochem. Soc.*, 1951, **98**, 207.
5. Olivier-Bourbigou H. and Magna L. *J. Mol. Catal. A Chemical* 2002, **182**, 419.
6. Hussey C. L. In *Advances in Molten Salts Chemistry*, Mamantov G. and Mamantov C. (eds), Elsevier, New York, 1983, Vol. 5, p. 185.
7. Øye H. A., Jagtoyen M., Oksefjell T. and Wilkes J. S. *Mater. Sci. Forum* 1991, 73–75, **183**.
8. Suarez P. A. Z., Einloft S., Dullius J. E. L., de Souza R. F. and Dupont J. *J. Chim. Phys.* 1998, **95**, 1626.
9. Seddon K. R., Stark A. and Torres M.-J. *Pure Appl. Chem.* 2000, **72**, 2275.
10. Wilkes S. J. and Zaworotko M. *Chem. Commun.* 1990, 965.
11. Fuller J., Carlin R. T., De Long H. C. and Haworth D. *Chem. Commun.* 1994, 299.
12. Olivier H. and Favre F. United States Patent 6,245,918, 2001.
13. Bonhôte P., Dias A.-P., Papageoriou N., Kalyanasundaram K. and Grätzel M. *Inorg. Chem.* 1996, **35**, 1168.
14. Holbrey J. D. and Seddon K. R. *J. Chem. Soc., Dalton Trans.* 1999, 2133.
15. Yoshizawa M., Hirao M., Ito-Akita K. and Ohno H. *J. Mater. Chem.* 2001, **11**, 1057.
16. Carmichael A. J. and Seddon K. R. *J. Phys. Org. Chem.* 2000, **13**, 591.
17. (a) Fletcher K. A., Storey I. A., Hendricks A. E. and Pandey S. *Green Chem.* 2001, **3**, 210; (b) Dzyuba S. V. and Bartsch R. A. *Tetrahedron Lett.* 2002, **43**, 4657; (c) Aggarwal A., Lancaster N. L., Sethi A. R. and Welton T. *Green Chem.* 2002, **4**, 512.
18. Armstrong D. W., He L. F. and Liu Y.-S. *Anal. Chem.* 1999, **71**, 3873.

19. (a) Dymek Jr C. J., Grossie D. A., Fratini A. V. and Adams W. W. *J. Mol. Struct.* 1989, **213**, 25; (b) Elaiwi A., Hitchcock P. B., Seddon K. R., Srinivasan N., Tan Y.-M., Welton T. and Zora J. A. *J. Chem. Soc., Dalton Trans.* 1995, 3467; (c) Takahashi S., Suzuya K., Kohara S., Koura N., Curtiss L. A. and Saboungi M.-L. *Z. Phys. Chem.* 1999, **209**, 209; (d) Larsen A. S., Holbrey J. D., Tham F. S. and Reed C. A. *J. Am. Chem. Soc.* 2000, **122**, 7264.
20. Huang J.-F., Chen P.-Y., Sun I.-W. and Wang S. P. *Inorg. Chim. Acta* 2001, **320**, 7.
21. (a) Anthony J. L., Maginn E. J. and Brennecke J. F. *J. Phys. Chem. B* 2002, **106**, 7315; (b) Dyson P. J., Laurenczy G., Vallance J. and Welton T. in preparation.
22. (a) Sheldon R. A. *Chem. Commun.* 2001, 2399; (b) Dupont J., de Souza R. F. and Suarez P. A. Z. *Chem. Rev.* 2002, **102**, 3667–.
23. Brown R. J. C., Dyson P. J., Ellis D. J. and Welton T. *Chem. Commun.* 2001, 1862.
24. Joó F. *Aqueous Organometallic Catalysis*, Kluwer, Dordrecht 2001.
25. (a) Monteiro A. L., Zinn F. K., de Souza R. F. and Dupont J. *Tetrahedron Asymm.* 1997, **8**, 177; (b) Berger A., de Souza R. F., Delgado M. R. and Dupont J. *Tetrahedron Asymm.* 2001, **12**, 1825; (c) Brown R. A., Pollet P., McKoon E., Eckert C. A., Liotta C. L. and Jessop P. G. *J. Am. Chem. Soc.* 2001, **123**, 1254.
26. Wasserscheid P., Bosmann A. and Bolm C. *Chem. Commun.* 2002, 200.
27. Dupont J., Fonseca G. S., Umpierre A. P., Fichtner P. F. P. and Teixeira S. R. *J. Am. Chem. Soc.* 2002, **124**, 4228.
28. Sheldon R. A., Madeira Lau R., Sorgedrager M. J., van Rantwijk F. and Seddon K. R. *Green Chem.* 2002, **4**, 147.
29. Scurto A. M., Aki S. N. V. K. and Brennecke J. F. *J. Am. Chem. Soc.* 2002, **124**, 10276.

5 Reactions in Water

In one way it seems strange to have a chapter on water in a book on alternative solvents. For many hundreds of years water was the only solvent available to chemists to carry out their reactions. It was not until organic solvents came into use that a whole new era of chemistry was born, and many types of reactions were conducted and compounds made that previously had not been thought possible. Some of the reactions that were only considered possible in organic solvents are now being conducted using water as a solvent, and this is very much at the forefront of solvent replacement research [1]. In this chapter, we overview the unique properties of liquid water, and then discuss its application as a solvent for organic chemistry. This is followed by an overview of the use of water in multiphasic systems, including aqueous-organic reactions and phase transfer catalysis.

5.1 THE STRUCTURE AND PROPERTIES OF WATER

Water is the most abundant solvent and in fact the most common molecule on the planet and the third most abundant in the universe, following H_2 and CO. It is the only naturally occurring inorganic compound that is a liquid at room temperature (elemental mercury is also liquid at room temperature and occasionally occurs in nature in the uncombined state). It is an excellent solvent capable of dissolving many ionic compounds as well as some covalent molecules. The high polarity of water enables it to solvate ions efficiently and any ion dissolved in water is associated with several water molecules. Water is a good HBD and HBA, and has a high dielectric constant. This reduces the strength of the electrostatic forces between dissolved ions and allows them to separate and move freely in solution. The key physical properties of water are listed in Table 5.1.

5.1.1 The Structure of Water

Water is often referred to as anomalous in its behaviour, and its properties are dominated by its ability to form H-bonds. It has high boiling point, melting point,

Chemistry in Alternative Reaction Media   D. Adams, P. Dyson and S. Tavener
© 2004 John Wiley & Sons, Ltd   ISBNs: 0-471-49848-3 (Cloth); 0-471-49849-1 (Paper)

Table 5.1 Physical properties of water

Melting point	$0\,°C$
Boiling point	$100\,°C$
Triple point	$0.01\,°C$
Critical temperature and pressure	$374\,°C$, $22.1\,MPa$
Density (at $4\,°C$)	$1.00\,g\,cm^{-3}$
Latent heat of vaporization	$2.26\,kJ\,g^{-1}\,K^{-1}$
Specific heat capacity	$4.19\,J\,g^{-1}\,K^{-1}$
Cohesive energy density	$2302\,MPa$
Internal pressure	$151\,MPa$
Hildebrand solubility parameter	$47.9\,MPa^{0.5}$
Dielectric constant	78.30
Dipole moment	$5.9 \times 10^{-30}\,C\,m$
E_T^N	1.000 (defined)
α	1.17
β	0.47
π^*	1.09
Donor number	1.46
Acceptor number	54.8

critical temperature ($374\,°C$, cf. $31.1\,°C$ for CO_2), surface tension and viscosity, all of which are influenced by the ability of each water molecule to form up to four H-bonds (two as a donor and two as acceptor). The intermolecular forces in water are much stronger than in other solvents. Unlike most compounds, water undergoes an increase in volume on freezing, and displays maximum density at $4\,°C$ (i.e. above its melting point). This latter anomaly is explained by the fact that ice has a rigid but open structure in which all the molecules are tetrahedrally coordinated and involved in four straight H-bonds, as shown in Figure 5.1. On melting, some of the H-bonds break and others bend allowing the structure to collapse into a more closed form. Thus two different, opposing processes are at work: thermal bond breaking leading to an increase in density, and thermal expansion which decreases density.

The simple chemical formula, H_2O, gives little indication of the sheer complexity of the structure of liquid water. Investigation of water using X-ray techniques, and subsequent data modelling, suggests that water has medium range structure and contains structured regions with regular, ice-like lattice, chains, clusters and other polymeric types, random regions with little or no structure, and single, mobile, water molecules which may permeate the structured regions, as depicted in Figure 5.2. Likewise, the ordered regions may contain holes, vacancies and faults in the structure.

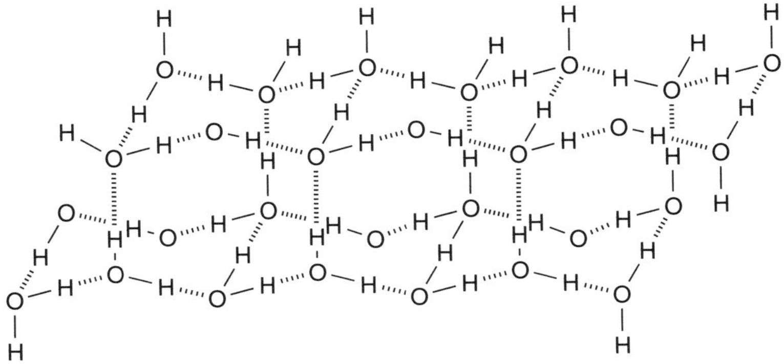

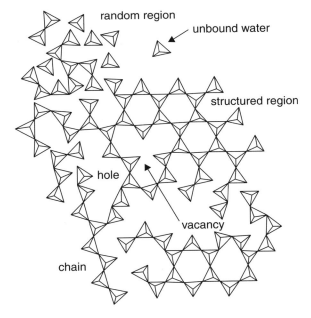

Figure 5.1   Hexagonal ice structure showing open, tetrahedrally coordinated structure

Figure 5.2   Two-dimensional representation of some of the structures present in liquid water

Recent interpretations of diffraction data suggest that water may contain regular clusters of as many as 280 molecules [2]. These large clusters have icosohedral symmetry and adopt two forms, expanded and compressed, (the latter appearing somewhat similar to a deflating soccer ball), which are thought to exist in a temperature and pressure dependent equilibrium.

Polywater

New claims for the structure of water are usually met with much caution and even a little cynicism. The difficulty in determining the structure of even a 'simple' liquid such as water is highlighted by the infamous *polywater* fiasco of the 1960s and early 1970s, which is still frequently sited (alongside cold fusion) as a cautionary tale for scientists. Soviet chemists had reported that by careful purification and condensation of water through a glass capillary, they had prepared small quantities of a new polymorph of water [3], with a viscosity 15 times that of 'normal' water, and an elevated boiling point of up to 600 °C. Polywater (which was sometimes referred to as 'anomalous water') was believed to contain very strong, symmetrical H-bonds and existed as hexagonal rings or polyhedral clusters [4]. Many other scientists took to their laboratories to investigate this phenomenon, and a large amount of research ensued. It was even feared that if polywater escaped into the wild it would seed the oceans and destroy life as we know it [5]. After five years of controversy and irreproducible results, it was found that the polywater was a solution of impurities including grease, human sweat and dissolved silica.

As well as being a good HBD and HBA, water is truly amphoteric and partly dissociated in solution, giving rise to tiny concentrations of $[H_3O]^+$ and $[OH]^-$, as shown in Equation 5.1. The equilibrium is influenced by the presence of solutes in the water.

$$2H_2O \rightleftharpoons H_3O^+ + OH^- \quad K = 1 \times 10^{-14} \tag{5.1}$$

While NaCl (rocksalt or common salt) exists as an equal number of hydrated Na^+ and Cl^- ions in aqueous solution, such a simple picture is only appropriate for ions of low charge. Ions with a higher charge density, such as Al^{3+} and Sn^{4+}, polarize the surrounding water molecules and do not form simple hydrated species. Instead, they polarize the water to the point of dissociation, forming hydroxyl complexes such as $Al(OH)_3$ and $Sn(OH)_4$ which are insoluble in neutral water, but become increasingly soluble through reaction as the acidity or basicity of the solution increases. Small hard metal cations that polarize water will exhibit acidic behaviour when present in water. Similarly, anions are also hydrated in aqueous solution, and small anions are sufficiently polarizing to produce an active supply of hydroxyl anions in solution (Figure 5.3). This is the reason why some attempted nucleophilic substitutions are unsuccessful (reactions involving fluoride are notoriously difficult) in the presence of water: the anion is substantially hydrated and polarizing, thus hydroxide is the active nucleophile.

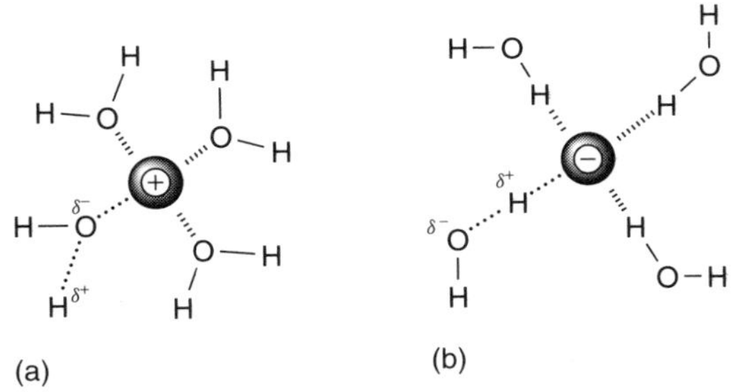

Figure 5.3 Polarization of water molecules by (a) a cation and (b) an anion

5.1.2 Near-Critical Water

Water superheated to between 200 and 374 °C at sufficient pressure to keep it liquid is known as *near-critical water* (NCW) or *sub-critical water*. Under these conditions, the H-bond network is partly broken down and the polarity and density decrease. When near-critical, water behaves more like an organic solvent. It has been used as a solvent for organic synthetic chemistry [6], and its solvatochromic parameters, E_T^N, α and β have been reported [7]. Increasing the temperature reduces both E_T^N (a general measure of solvent polarity), and α, which is an indication of HBD ability, as shown in Figure 5.4. However, β, the HBA ability, remains relatively constant throughout the temperature range. At 300 °C, E_T^N and α decreased to levels which are comparable to that of ethanol at ambient temperature and pressure (ethanol has values of 0.86 and 0.654 for α and E_T^N, respectively). Water becomes completely miscible with toluene at 308 °C and 22 MPa [8] and NCW has been used as a reaction solvent for Friedel–Crafts acylations of phenol and *p*-cresol using acetic acid [9].

5.1.3 The Hydrophobic Effect

The term *hydrophobic* is commonly used to describe the apparent immiscibility observed for many organic compounds in water. The name itself is misleading in terms of describing the interaction between the water and the organic molecules. Hydrophobic translates from Greek as 'water hating', implying that the organic molecules are repelled by the water molecules. This is not the case: the induced dipolar interactions between a hydrocarbon and the dipolar water are in fact much stronger than the dispersive forces that exist between the hydrocarbon molecules themselves. Observe how petrol will spread out over the surface of a puddle to maximize its contact with the water. The enthalpy of dissolution of a hydrocarbon in water is exothermic ($\Delta H < 0$) so the observed hydrophobic effect must be caused by an entropic factor. Water itself has a very large cohesive energy density (CED = 2302 MPa, cf. cyclohexane = 285 MPa), and it is this

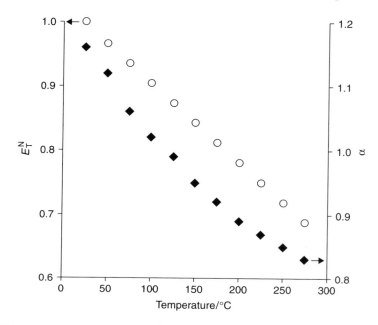

Figure 5.4   Effect of temperature on E_T^N and α for sub-critical water

which is responsible for the hydrophobic effect. In order to minimize disruption to the H-bond network, water restructures itself around dissolved hydrocarbon molecules leading to restricted motion, an increase in order and a decrease in entropy. As $\Delta G = \Delta H - T\Delta S$, this leads to an increase in the free energy of the system. Aggregation of organic molecules will decrease the number of water molecules that need to undergo this restructuring (Figure 5.5) and reduces the effect. The organic molecules congregate and two phases are formed.

This aggregation of organic molecules in water actually leads to accelerated rates for some organic reactions in water. Diels–Alder and other sigmatropic reactions work very well in water, despite not being soluble or miscible in that solvent. This effect is discussed further in Chapter 7.

5.1.4   The Salt Effect

The dissolution of electrolytes in water has a strong effect on the internal pressure of the solvent, a phenomenon known as the *salt effect*. Almost all electrolytes (perchloric acid is the exception) increase the internal pressure of water by *electrostriction*, a term used to describe the polarization and attraction of water molecules. The effect of this internal pressure is to 'squeeze out' the organic

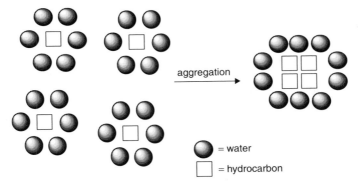

Figure 5.5 The hydrophobic effect. Aggregation of hydrocarbon molecules in water reduces the number of molecules with restricted motion

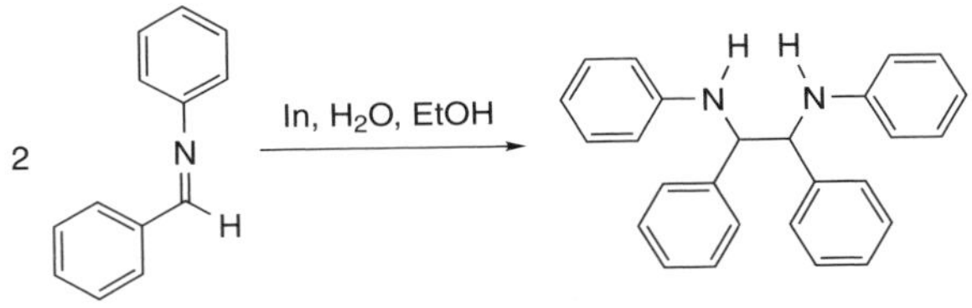

Scheme 5.1 Indium mediated imine coupling

molecules, reducing their solubility and increasing the hydrophobic effect. The internal pressure of the water acts on an organic molecule in the same way that an external pressure would, and an increase on the internal pressure has an accelerating effect on reactions with significant volumes of activation ($\Delta V^{\dagger}$). For example, Diels–Alder reactions will proceed twice as fast in 4.86 M LiCl solution than they will in water alone. This is discussed further in the case study in Chapter 7. Another reaction displaying a salt effect is the indium metal mediated imine coupling reaction shown in Scheme 5.1 [10]. This reaction is accelerated by the addition of ammonium chloride, which increases the internal pressure of the water, and evidently the reaction has a negative volume of activation.

5.2 THE BENEFITS AND PROBLEMS ASSOCIATED WITH USING WATER IN CHEMICAL SYNTHESIS

Water is abundant and has many properties that make it a desirable solvent (perhaps only superseded by not using a solvent at all). These desirable properties include the following:

1 Polar and therefore relatively easy to separate from apolar solvents.
2 Nonflammable and incombustible.

3 Cheap, and widely available.
4 Odourless and colourless–contamination is therefore easy to recognize.
5 Density is sufficiently different from most organic substances to allow convenient separation.
6 High thermal conductivity, heat capacity and heat of evaporation. High specific heat capacity means that exothermic reactions may be controlled effectively.

Enzyme Catalysed Reactions in Water and Aqueous Biphase

Biochemical reactions take place inside cells in the presence of water, and it might therefore by expected that water should be a useful solvent for enzyme-catalysed reactions. Many enzymatic reactions also perform well in biphasic systems–some enzymes, including lipases, possess a hydrophobic pocket or 'lid', which protects the active site and is closed in aqueous solution [12]. This lid will change its structure on contact with a droplet of an organic solvent or the pure substrate to minimize the hydrophobic interactions, opening the lid and allowing the reaction to proceed. Enzymatic reactions will proceed in a wide range of other reaction media, including ionic liquids and supercritical CO_2 [13]. When an organic solvent is used, it is best to have at least some water present, as this helps the enzyme hold its shape. Biocatalysts are generally most active either in water, or in a solvent which displays a high value for the 1-octanol/water partition coefficient (see Chapter 2), in which any water in the system forms an aqueous shell around the enzyme [14]. Polar solvents such as nitromethane and acetonitrile are thought to strip away this shell and reduce enzyme activity and selectivity. For example, the enzyme *substilisin Carlsberg* catalyses enantioselective transesterification reactions, and is much more selective in less polar solvents. Note that a small amount of water must be present for the reaction to proceed efficiently.

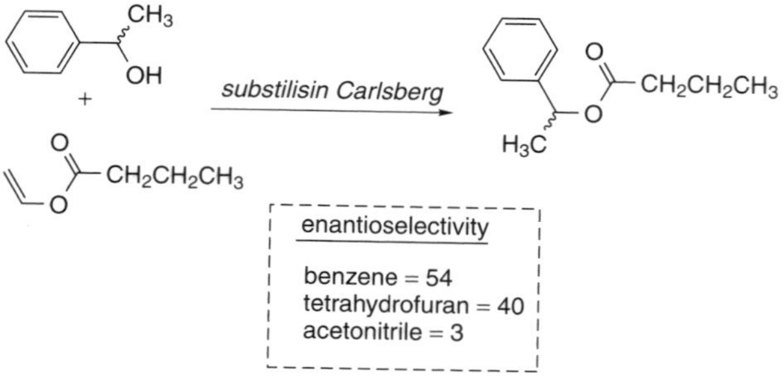

enantioselectivity
benzene = 54
tetrahydrofuran = 40
acetonitrile = 3

Scheme 5.2

Since there are so many benefits to using water as a solvent to conduct chemical reactions it is perhaps surprising that it is not used more frequently. Some of the reasons why water is not used more often are:

1 Not all reagents dissolve in water (although this is not always a disadvantage).
2 Some compounds react in an adverse way with water and many catalysts are deactivated by reaction with water.
3 Water may be difficult to purify after reaction.
4 High specific heat capacity means that distillation requires high energy, and is difficult to heat or cool rapidly [11].

5.3 ORGANOMETALLIC REACTIONS IN WATER

Many chemists have unpleasant memories of attempting to prepare organomagnesium Grignard reagents as an undergraduate practical experiment, and recall how critical it is to ensure that the system is clean and dry to avoid hydrolysis of the reagent. On this basis it might therefore be predicted that reactions involving active organometallic reagents would be impossible in water. This is not the case, and there are many documented examples of organic syntheses using organometallic reagents that have been effectively carried out in water. Reactions mediated by tin, zinc, lead, cadmium, mercury and bismuth have been performed successfully, and indium has been the subject of special interest. Indium has a very low first ionization energy, but does not react with boiling water and does not oxidize readily in air: it is these properties that make it useful for organometallic chemistry in water [15]. Using indium powder in water at room temperature, it is possible to alkylate at a ketone function in the presence of a delicate acetal function that would be hydrolysed in the presence of a more polarizing metal such as zinc (Scheme 5.3).

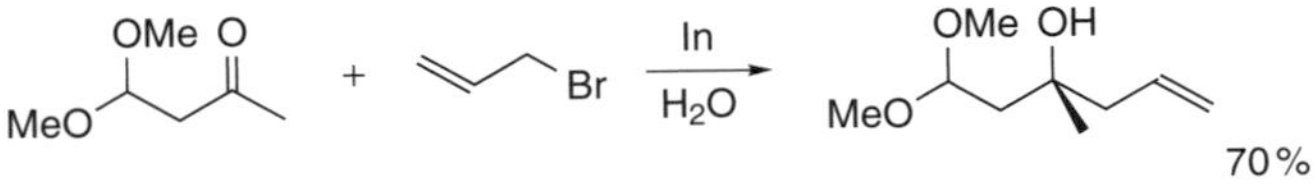

Scheme 5.3 Ketone alkylation promoted by indium in water

5.4 AQUEOUS BIPHASIC CATALYSIS

Water is particularly suitable for use in biphasic catalysis. It readily separates from organic solvents because of its polarity, density and because of the hydrophobic effect. Water will form biphasic systems with fluorous solvents, some ionic liquids, many volatile organic solvents, and also with $scCO_2$ [18].

Microwave Activation of Aqueous Reactions

The use of microwave radiation as a way of accelerating chemical reactions has been extensively investigated, and works best in solvents such as acetonitrile and water, which have high dielectric constants. There has been some dispute over whether or not a 'microwave effect' truly exists: in most cases where rigorous temperature control and kinetic measurements have been made, it appears that the rate of microwave and conventionally heated reactions are identical. However, microwave irradiation allows for extremely high heating rates without the thermal lag of a conventional conduction-heated device. If care is taken to cool the reaction rapidly, this rapid heating can allow the synthesis of difficult chemical target molecules which have limited thermal stability but which also require high activation energies. For example, the synthesis of heat sensitive aryl vinyl ketones may be carried out in quantitative yield by elimination of the corresponding quaternary ammonium salts under microwave conditions in the presence of water [16]. It seems likely that the cationic nitrogen centre is hydrated and so a localized heating effect occurs at the active centre.

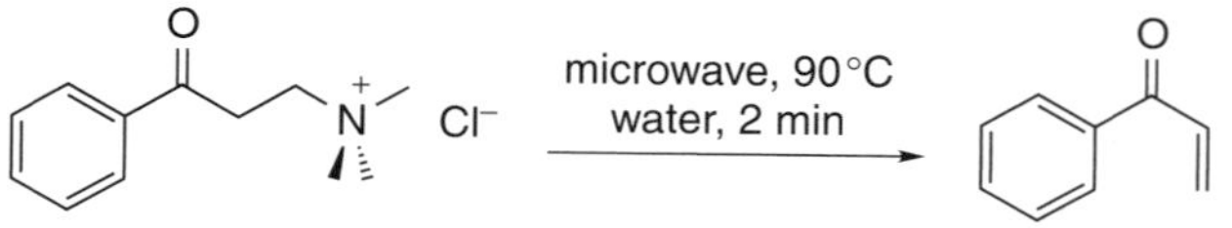

Scheme 5.4

Microwave activation is thought to be particularly applicable to reactions in water because it is tuned to water. It can therefore be used to selectively heat the water of hydration around a nucleophilic anion or other reacting species [17].

5.4.1 Ligands for Aqueous–Organic Biphasic Catalysis

A large number of ligands have been developed for use in aqueous–organic biphasic processes. They all share a common feature, this being the presence of one or more hydrophilic groups, which determine their solubility in water. However, once coordinated to a metal complex the solubility may change depending on the charge of the complex as a whole and the nature of the other ligands present. The most widely used group of ligands used in aqueous–organic biphasic catalysis are the sulfonated phosphines and some examples are illustrated in Figure 5.6.

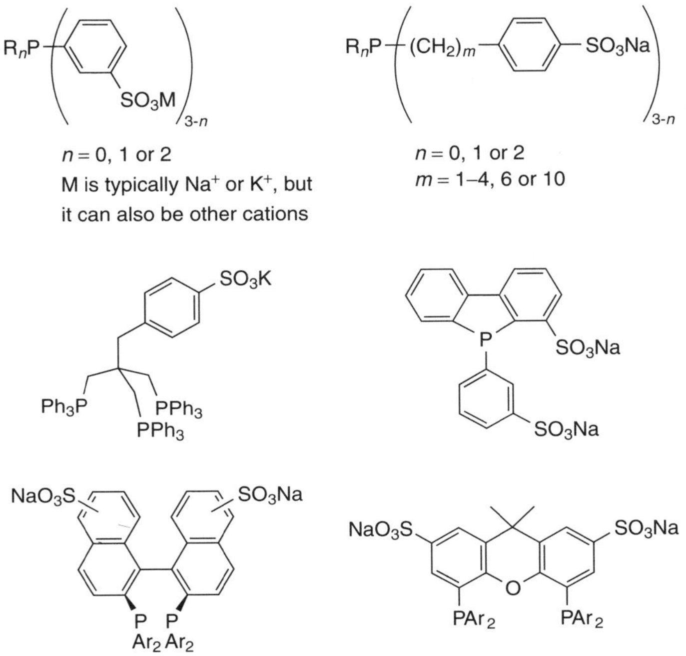

Figure 5.6 Some examples of sulfonated phosphine ligands widely used in aqueous–organic biphasic catalysis [19]

The solubility of these phosphine ligands in water can be extraordinarily high. The sodium salt of the *tris*-sulfonated triphenylphosphine (tppts) has a solubility in water of $1.1 \, kg \, l^{-1}$ [20]. Apart from high solubility, these sulfonated phosphines are widely used for several reasons including:

1   High solubility over a wide pH range.
2   Very poor solubility in nonpolar organic solvents.
3   They are usually stable under catalytic conditions.
4   Many can be prepared quite easily from currently available phosphines by derivatization of a phenyl ring with the sulfonate group.

With this last point in mind, the synthesis of *mono-*, *bis-* and *tris*-sulfonated triphenylphosphine will be described, but the same methodology can be applied to the preparation of other sulfonated arylphosphines. In general, these phosphines are made by direct sulfonation using fuming sulfuric acid (oleum) [21]. The extent of sulfonation is determined by the SO_3 strength, as well as factors such as the temperature and time of the reaction. The monosulfonated phosphine (tppms) is prepared using oleum of 20 % SO_3 strength, with typically 30 % SO_3 used for higher levels of substitution [22]. The main problem with this reagent is that mixtures of products are often made which require extensive purification. Cleaner products are generally obtained by using anhydrous sulfuric acid in combination with orthoboric acid, which minimizes the extent of side reactions [23].

There are many nonsulfonated phosphines that induce high solubility in water. Such ligands contain groups that can form strong H-bonds and usually contain several nitrogen or oxygen atoms and even groups common in biological systems such as carbohydrates. Figure 5.7 shows a selection of the more widely used phosphine ligands of this type.

A wide range of polar, hydrophilic functional groups may be used to induce solubility in water, and these include sulfonates, alcohols, nitriles, amines,

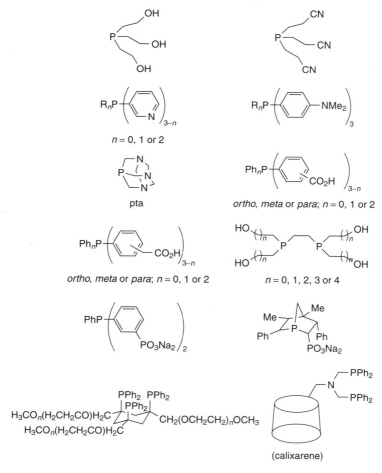

Figure 5.7 Some water-soluble phosphines [24]

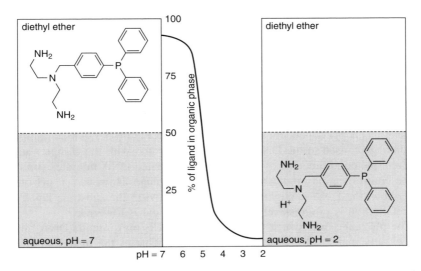

Figure 5.8 The extraction curve for the N3P ligand between water and diethyl ether

carboxylic acids, polyethylene glycols, phosphates and calixerenes. Some of these functional groups may react under catalytic conditions and care must be taken to select an appropriate phosphine ligand that will have sufficient stability for the required task. In certain cases reaction of the ligand can be used to an advantage. For example, basic nitrogen atoms are protonated or deprotonated depending upon the pH of the solution [25]. This property can be put to good use in catalysis. In a biphasic process in which the catalyst is immobilized in water it may be desirable at some stage to remove the catalyst. Distillation is not only expensive, but it can also cause the decomposition of the catalyst. A far more effective method is to modify the pH to alter the solubility properties of the catalyst and this is exactly what happens with N-containing ligands. At high pH the ligand is deprotonated and preferentially soluble in the organic phase in a biphase mixture. As the pH is lowered, protonation of the ligand takes place and the catalyst becomes preferentially soluble in the aqueous phase. This process occurs for the ligand N3P, and the effect of pH on the partitioning of this ligand is illustrated in Figure 5.8.

Attaching a pyridine group on the phosphine (e.g. PPh$_2$Py) is also believed to enhance the catalytic activity in methyl methacrylate synthesis by picking up protons in solution and transferring them to the metal centre, which then uses them in the reaction [26]. Such a process has been termed a *proton messenger*.

Apart from phosphine ligands, a number of other types of ligands have been modified to induce water solubility, the most widely used ones being porphyrin based ligands such as those shown in Figure 5.10. The same water solubilizing groups used in phosphines are used to induce hydrophilicity in these ligands.

Medical Applications of pH-Sensitive Organometallic Catalysts

The ruthenium complex [Ru(η^6-p-cymene)Cl$_2$(pta)] contains the aliphatic phosphine 1,3,5-triaza-7-phosphatricyclo[3.3.1.1]decane (pta, see Figure 5.7). This complex has pH dependent cytotoxicity, which leads to selective inhibition of tumors [27]. Healthy cells have a pH of ca. 7.2 and many rapidly growing cancer cells have a pH of ca. 6.8. DNA damage induced by [Ru(η^6-p-cymene)Cl$_2$ is pH dependent as can be seen from Figure 5.9, which compares the damage caused to the DNA when it is incubated with the complex under different pH conditions. Above pH 7.0, the incubated DNA migrates similarly to the control sample, indicating little or no damage. However, at pH of 7.0 and below, the DNA is slightly retarded compared to the substrate DNA and the retardation is progressively increased as the pH is lowered.

The pH range over which [Ru(η^6-p-cymene)Cl$_2$(pta)] retards DNA migration closely matches the pK_a of the pta ligand. One of the ligand's nitrogen atoms becomes protonated at low pH and, as DNA is negatively charged, it is expected that the interaction between these two species would be promoted as they each carry opposite charges. The importance of this result is that DNA binding is not favoured at physiological pH. However, many diseased cells have a reduced pH, typically 6.8, due to metabolic changes in part associated with the accelerated cell division [28]. Thus, [Ru(η^6-p-cymene)Cl$_2$(pta)] has a higher affinity for DNA in cancer cells, compared to healthy cells, providing a means of targeting the cancer cells without damaging healthy cells.

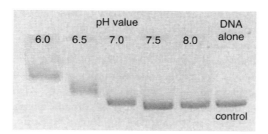

Figure 5.9 Comparison of DNA damage induced by [Ru(η^6-p-cymene)Cl$_2$(pta)] incubated at different pH values; shown by DNA migration in an agarose gel. Reproduced by Permission of the Royal Society of Chemistry

The water-soluble ligands described above, together with many others, are used to conduct a wide range of catalytic reactions in water. These reactions include hydrogenation, hydroformylation, oxidation, C–C coupling and polymerization reactions [30]. Many of these reactions are discussed in detail in Chapters 7–11.

The immiscibility of water with organic substrates may often be put to good use by adopting a biphasic approach, and thus improving the separation and

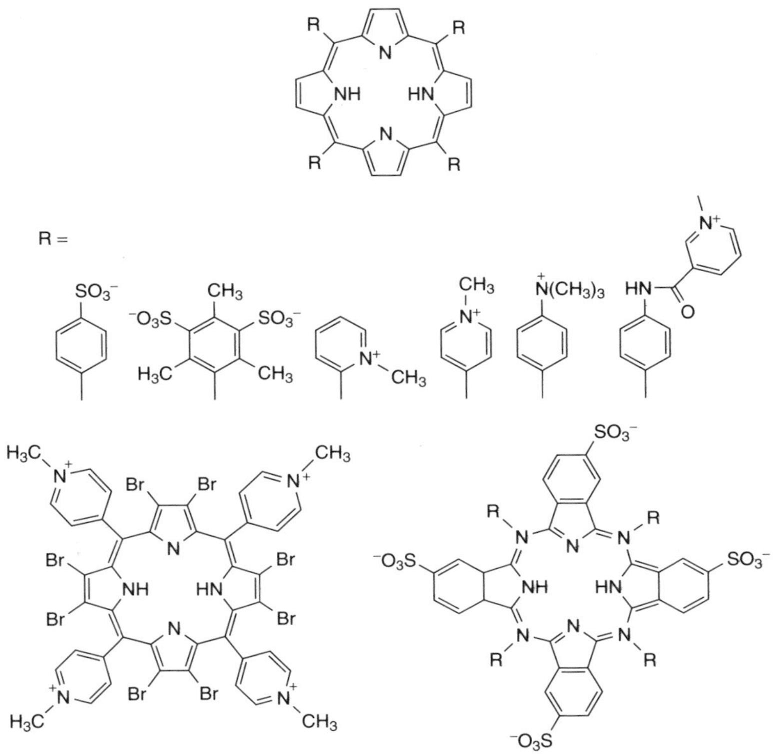

Figure 5.10 Some examples of water-soluble porphyrin ligands [29]

recovery of products and catalysts. However, the poor solubility of certain organic substrates in water may reduce reaction rates, and therefore limit the effectiveness of water as a reaction solvent. A complementary approach is the use of *phase transfer catalysis* (PTC) [31]. This differs from the aqueous biphase reactions discussed above, in that, in a phase transfer reaction, a polar reagent is usually made soluble in, or available to, an organic substrate or solvent, whereas in an aqueous biphase the catalyst or substrate is made soluble in the aqueous phase. Phase transfer is particularly useful for bringing together reactants with markedly different natures or polarities, for example an organic substrate and an inorganic salt, without the need for a mutual solvent. PTC and some related systems will be described in the following sections.

5.5 PHASE TRANSFER CATALYSIS

A phase transfer catalyst can be defined as a substance that will increase the rate of reaction between substrates present in separate phases. Phase transfer

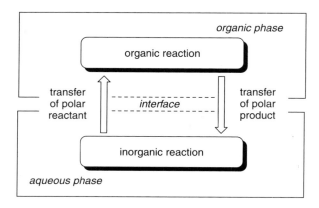

Figure 5.11   A simplified representation of the steps in a phase transfer reaction cycle

is perhaps not catalysis in the conventional sense: it often does not affect the intimate mechanism of the reaction. It does, however, speed up a reaction by ensuring a ready supply of a necessary reagent to the phase in which the reaction is occurring (see Figure 5.11).

The vast majority of reported phase transfer catalysts may be divided into two types; those which have an organophilic cation, usually centring on a nitrogen or phosphorus atom, and those which are polydentate complexing agents. Quaternary ammonium and phosphonium salts (often referred to as *quats* or *oniums*) fall into the former category, and crown ethers, polyethylene glycols (PEGs) and cryptands belong to the latter (Figure 5.12). More sophisticated phase transfer catalysts have been synthesized, such as silacrowns [32], and onium salts with crown ether groups at the ends of their alkyl chains [33]. However, as PTC is advantageous from an economic point of view, the simpler, more readily available quats are by far the most popular.

Phase transfer agents can catalyse a variety of organic reactions, including simple nucleophilic substitution [34], aromatic halogen exchange (Halex), Friedel–Crafts reactions, Wittig reactions [35], oxidations, and the generation of carbenes [36]. Catalytic systems have been devised which can transfer inorganic anions (including hydroxides) into organic phases, organic salts into aqueous phases, and neutral polar molecules such as hydrogen peroxide can be extracted into organic solvents. Phase transfer can occur between two immiscible liquids, a solid and a liquid, or even between a liquid and a gas [37]. Nonpolar organic solvents may be used, and in some cases the need for an organic solvent is eliminated completely, i.e. the reaction is conducted using the neat substrate as the organic phase [38]. Clearly this versatility provides many opportunities for the application of PTC, some examples of which have included part of the synthesis of a naturally occurring antibiotic [39], and the use of tetrabutylammonium chloride to transfer perchlorate anions into organic phases for the decontamination of

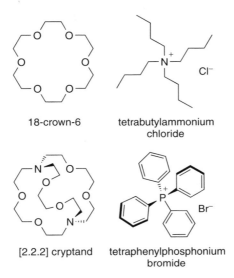

18-crown-6   tetrabutylammonium
                   chloride

[2.2.2] cryptand   tetraphenylphosphonium
                         bromide

Figure 5.12 Typical phase transfer catalysts

chemical warfare agents [40]. PTC is at its most effective when used to provide a supply of anionic or uncharged reactants to an organic phase. Typically, a PTC reaction will use less than 5 mol% of catalyst, above which quantity the cost effectiveness of the system decreases.

There are a number of advantages to be gained in using a phase transfer system to conduct a reaction, including:

1 Elimination of the need for organic solvents; may often be performed in just water and the neat substrate.
2 Inexpensive inorganic reagents may be used.
3 Separation may be improved.
4 Reaction temperatures (and therefore rates) may be controlled by the rate of stirring.
5 Higher productivity due to increased reaction rates.
6 Better selectivity due to lower operating temperatures.

However, there are a number of drawbacks associated with PTC, including:

1 Requires a catalyst, which may be toxic or expensive.
2 Emulsions may be formed, making separation difficult.
3 Catalyst may be difficult to recover.
4 Generally requires vigorous mixing.
5 Waste water may be difficult to purify sufficiently if it is contaminated with soluble organic compounds.

5.5.1   The Transfer of Nucleophiles into Organic Solvents

Probably the most important group of phase transfer reactions, and certainly the commonest, are those in which an anion is transferred from the aqueous phase into the organic solvent, where nucleophilic substitution occurs. These would once have been performed in a dipolar aprotic solvent such as DMF. A good example is the reaction between an alkyl halide (such as 1-chlorooctane), and aqueous sodium cyanide, shown in Scheme 5.5. Without PTC, the biphasic mixture can be stirred and heated together for 2 weeks and the only observable reaction will be hydrolysis of the cyanide group. Addition of a catalytic amount of a quaternary onium salt, or a crown ether, however, will lead to the quantitative conversion to the nitrile within 2 h.

Displacement by cyanide works particularly well, and many other nucleophilic substitution reactions are enhanced by PTC. Most monovalent anions can be transferred, including alkoxides, phenoxides, thiocyanates, nitrates, nitrites, super-oxides and all of the halides. Divalent anions are usually too hydrophilic to be transferred into the organic phase.

Typically, the organic substrate in these reactions is a haloalkane. Primary haloalkanes will generally give 100 % substitution products, but tertiary and cyclohexyl halides usually undergo 100 % elimination, with secondary haloalkanes producing a mixture of the two. Studies of the chloride and bromide displacements of (R)-2-octyl methanesulfonate have shown that phase transfer displacements proceed with almost complete inversion of stereochemistry at the carbon centre, indicating an S_N2-like mechanistic pathway [41].

5.5.2   Mechanisms of Nucleophilic Substitutions Under Phase Transfer Conditions

The commonly used quaternary onium salts work by associating with the inorganic anion and transferring it into the organic phase. Complexing agents such

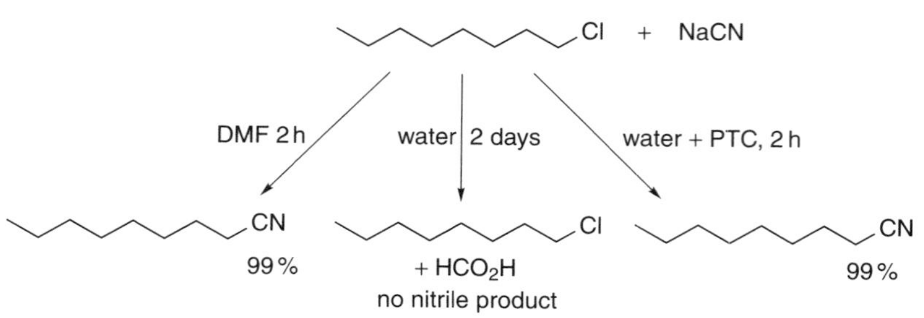

Scheme 5.5   The products of a simple nucleophilic substitution reaction is dependent on reaction conditions, and a PTC is essential for successful cyanation in water

Plate 1 (Figure 1.13) Solvatochromic behaviour of Reichardt's dye, in (from left to right) dichloromethane, acetone, acetonitrile, ethanol and methanol

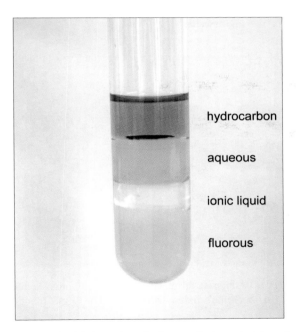

hydrocarbon

aqueous

ionic liquid

fluorous

Plate 2 (Figure 2.5) Four liquid phases in a single test tube (note that the coloured phases have been dyed for clarity) [photograph by DA; thanks to Glen Capper for supplying the ionic liquid]

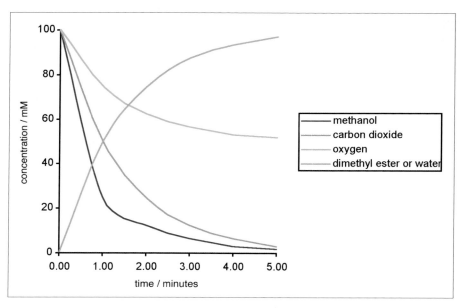

Plate 3 (Figure 2.8) Change in concentration of reactant and products for the reaction shown in Equation 2.5

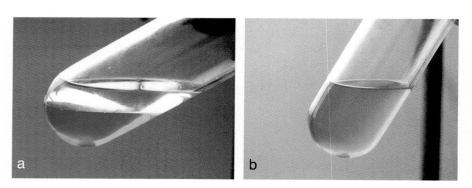

Plate 4 (Figure 3.2) (a) A biphase of toluene (top) and a fluorous-soluble catalyst in PP3 (bottom) at 25°C; (b) the same mixture when heated to 70°C [photographs by James Sherrington]

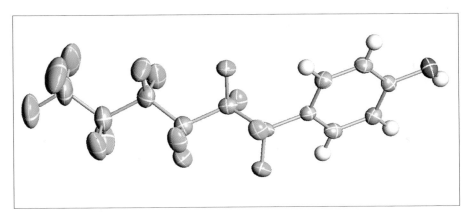

Plate 5 (Figure 3.5)   Crystal structure of 4-tridecafluorohexylphenol [24]

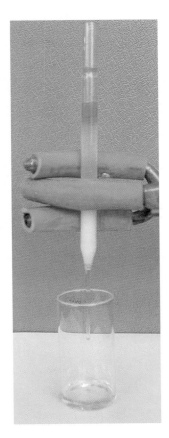

Plate 6 (Figure 3.7)   Separation of catalyst from products using a short column of FRPSG
[Photograph by Ben Croxtall]

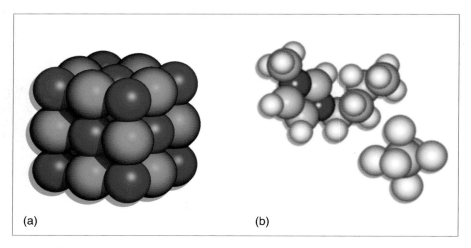

(a)   (b)

Plate 7 (Figure 4.1) Sodium chloride **(a)** can easily form a regular lattice through packing of the spherical anions and cations. It is much harder for [bmim][PF$_6$] **(b)** to do so because the charges are dispersed across the large ions, and the cation has low symmetry

Plate 8 (Figure 10.4) Polystyrene spheres prepared by emulsion polymerisation methods. Because they may be packed together to form columns or beds, these spheres find applications in separations, ion exchange, and as supports for catalysts [Photographs by John Olive]

as crown ethers, however, form strong interactions with the inorganic cation, and thus transfer the whole inorganic salt across the phase boundary, as shown in Figure 5.13.

There are two generally accepted mechanisms for simple phase transfer reactions under neutral conditions. The first of these is a mechanism in which the whole cation–anion complex moves between the two phases as shown in Scheme 5.6 [42].

However, the observation that very organophilic ammonium salts are also very active phase transfer agents suggests that the catalyst cation need not actually enter the aqueous phase, and the reaction can occur entirely at the interface, as depicted in Scheme 5.7 [43]. This process is known as the *interfacial mechanism.*

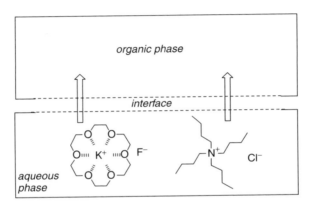

Figure 5.13 A representation of the differing actions of phase transfer agents. Organic cations will form an ion pair in order to transfer an anion, while complexing agents such as crown ethers transfer the whole inorganic salt across the phase boundary

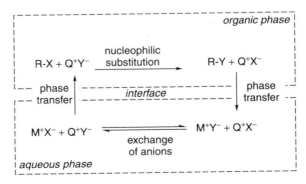

Scheme 5.6 Mechanism for biphasic anion displacement, in which the cation performing the phase transfer reaction, Q^+, shuttles between the two phases

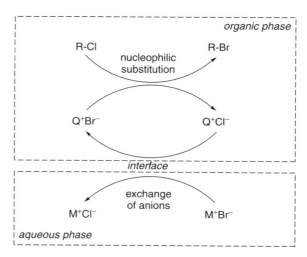

Scheme 5.7   Interfacial mechanism for simple anion displacement. In PTC, Q^+ is resident in the organic phase and draws the anion across the interface

The predominant mechanism in any particular system will be controlled by the lipophilicity of the catalyst, and by the polarity of the organic phase.

5.5.3   The Rates of Phase Transfer Reactions

Reaction rates of phase transfer reactions are sensitive to a number of factors, including the organic substrate and the inorganic salt, the catalyst itself, the organic solvent, and the amount of water used. Generally, if the rate of the reaction in the organic phase is significantly slower than the rate of phase transfer, there will always be sufficient supply of the transferred species and the reaction will follow its normal kinetic profile: typically *pseudo*-first order. These reactions have similar rate equations to homogeneous S_N2 reactions, but, because it is the catalyst quantity that limits the concentration of nucleophile in the organic phase, the rate equation is typically

$$\text{rate} = k[\text{substrate}][\text{catalyst}] \qquad (5.2)$$

For a single, homogeneous liquid phase the equation is:

$$\text{rate} = k[\text{substrate}][\text{nucleophile}] \qquad (5.3)$$

If the rate of phase transfer is much slower than the rate of reaction, then the reaction will be mass transport limited–typically *pseudo*-zeroth order, unless the availability of the nucleophile becomes significantly depleted over the course of the reaction [44]. If the rates of reaction and phase transfer are similar, then the kinetics may become complex. The rates of reactions in biphasic systems in general are discussed further in Chapter 2, and a detailed discussion of reaction rates in PTC reactions in particular can be found elsewhere [45].

5.5.3.1 The Influence of Catalyst Structure on Phase Transfer Reactions

The choice of the catalyst is an important factor in PTC. Very hydrophilic onium salts such as tetramethylammonium chloride are not particularly active phase transfer agents for nonpolar solvents, as they do not effectively partition themselves into the organic phase. Table 5.2 shows relative reaction rates for anion displacement reactions for a number of common phase transfer agents. From the table it is clear that the activities of phase transfer catalysts are reaction dependent. It is important to pick the best catalyst for the job in hand. The use of onium salts containing both long and very short alkyl chains, such as hexadecyltrimethylammonium bromide, will promote stable emulsions in some reaction systems, and thus these are poor catalysts.

Tetraalkylammonium and phosphonium salts are stable in acidic and neutral systems up to temperatures of between 100 and 150 °C, with phosphonium salts being more stable than the corresponding ammonium salts. For the same cation, stability decreases in the order $I^- > Br^- > Cl^-$. The stability is dramatically reduced in alkaline media, in which ammonium salts are more stable than phosphonium salts. More lipophilic onium salts often appear to be more effective phase transfer agents, but this may simply be due to the fact that they are in the right place to promote the reaction, namely the organic phase. This is demonstrated by the rates of reaction between n-octyl methanesulfonate and potassium bromide under water–chlorobenzene biphasic conditions, using a range of phase transfer catalysts. The *pseudo* first order rate constants, k_{obs}, vary by about two orders of magnitude. However, the second order rate constants, which take account of the proportion of catalyst in the organic phase, vary by only a factor of 2.5 (Table 5.3) [43].

Table 5.2 Relative reaction rates for various phase transfer catalysts in anion displacement reactions

Catalyst	Relative rates for PhS^-(aq) + 1-bromooctane	Relative rates for CN^-(aq) + 1-chlorooctane
n-Bu$_4$N$^+$ Br$^-$	1	1
n-Bu$_4$P$^+$ Cl$^-$	7	2.6
Ph$_4$P$^+$ Br$^-$	5.3	<0.05
Ph$_4$As$^+$ Cl$^-$	0.3	–
(n-Oct)$_3$N$^+$Me Br$^-$	5.9	1.8
C$_{16}$H$_{33}$N$^+$Et$_3$ Br$^-$	0.1	2

Source: (a) Herriott A. W. and Picker D. *J. Am. Chem. Soc.* 1975, **97**, 2345; (b) Starks C. M. and Liotta C. *Phase Transfer Catalysis: Principles and Techniques*, Academic Press, New York, 1978.

Table 5.3 The effect of catalyst structure and its organophilicity on the rate of reaction between n-octyl methanesulfonate and potassium bromide

Cation[a]	% in organic phase	$k_{obs}(10^{-5} \text{ s}^{-1})$	$k_2(10^{-5} \text{ mol}^{-1}\text{dm}^3 \text{ s}^{-1})$
$C_{16}H_{33}P^+Bu_3$	100	12.8	3.2
Bu_4P^+	97	10.4	2.7
Pr_4N^+	2.5	0.24	2.4
$PhCH_2N^+Pr_3$	17	1.3	1.9
Bu_4N^+	83	12.0	3.6
$PhCH_2N^+Bu_3$	95	8.2	2.1
Oct_4N^+	100	20.4	5.1
$C_{16}H_{33}N^+Pr_3$	94	15.9	4.0
$C_{16}H_{33}N^+Bu_3$	100	17.3	4.3

[a] All alkyl chains are linear.

In some reactions, the choice of PTC can actually alter the distribution of products; for example, the reaction of t-butyl acrylate with chloroform under basic phase transfer conditions gives two major products, as illustrated in Scheme 5.8, the amounts of which are catalyst dependent [46].

The change in selectivity on changing the catalyst is ascribed to lower lipophilicity and greater hydration of the tight Q^+ CCl_3^- ion pair, which reduces the activity of the trichloromethyl anion, allowing more time for decomposition to the carbene. The effect of catalyst structure may be summarized as follows [47]:

1 Large, soft, sterically hindered and delocalized cations such as Ph_4As^+ and $[Ph_3P=N=PPh_3]^+$ prefer to be associated with CCl_3^- anions.
2 Smaller, hard cations, such as alkyltrimethylammonium salts and crown ether–cation complexes, will promote the elimination of dichlorocarbene to form a hard ion pair of type Q^+Cl^-.
3 Intermediate phase transfer catalysts such as tetrabutylphosphonium and benzyltriethylammonium cations have no strong effect either way, and will generally give a mixture of products.

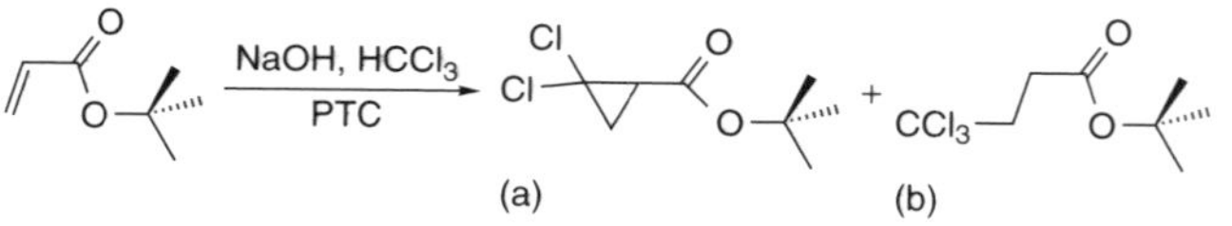

catalyst	yield (%)	
	(a)	(b)
$PhCH_2NEt_3Cl$	25	9
Me_4NBr	57	5

Scheme 5.8 Reaction of t-butyl acrylate with chloroform under basic phase transfer conditions

High Catalyst Quantities May Lead to the Formation of a Third Liquid Phase

The use of larger than normal quantities of catalyst leads to some surprising results. The elimination reaction of (2-bromoethyl)benzene in the presence of toluene and aqueous sodium hydroxide is catalysed by the presence of tetrabutylammonium bromide [48].

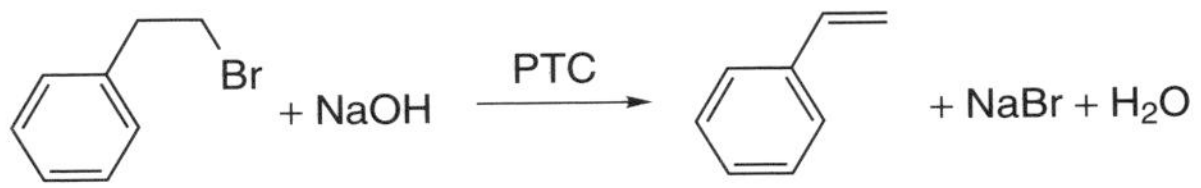

Scheme 5.9

The reaction proceeds as a normal two-phase PTC reaction with concentrations of catalyst below 5 mol%. As the concentration of the catalyst is increased to 11 mol%, a third liquid phase begins to form as droplets at the interface, resulting in a continuous phase at higher catalyst concentrations. The onset of this phenomenon is accompanied by a sudden five-fold increase in the reaction rate, although the rate is barely changed by further addition of catalyst. This third layer consists of (by weight) toluene (44 %), water (2 %) and tetrabutylammonium bromide (54 %). The phenomenon is not observed below 38 °C, and is cation specific. No third phase is formed at any temperature in similar reactions using tetrapropylammonium, tetrapentylammonium and tetrahexylammonium bromides. The same study showed that an increase in the concentration of hydroxide in the aqueous layer from 46 % to 49 % caused a drop in reaction rate by an order of magnitude, which is a rare example of a base catalysed reaction being inhibited by an excess of base. It is suggested that this increase in concentration causes dehydration of the third liquid phase with a consequent precipitation of catalyst and decrease in reaction rate.

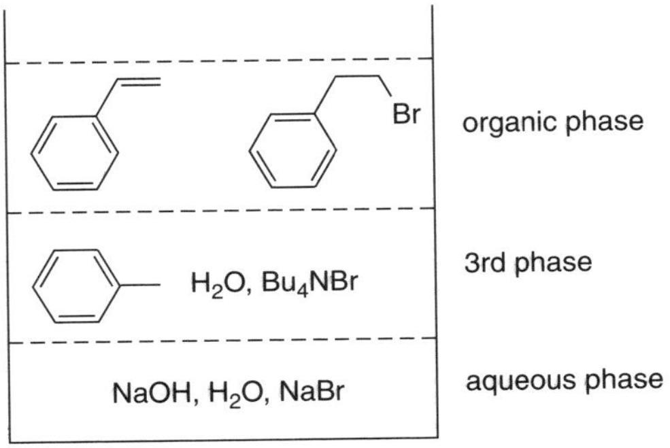

Scheme 5.10

5.5.3.2   The Effect of the Nucleophile on Phase Transfer Reactions

In general, for a successful PTC reaction, the attacking anion should be more lipophilic than the anion that is to be displaced. If this not the case, then a large excess of the inorganic salt is required to push the equilibrium towards the desired product.

A study of the anion displacement reactions of *n*-octyl methanesulfonate reveals striking differences in the reactivity of various anions under phase transfer and homogeneous conditions [43]. In the phase transfer catalysed reactions the reactivities of the anions were found to decrease in the order:

$$N_3^- > CN^- > Br^- \approx I^- > Cl^- > SCN^-$$

which deviates significantly from the known reactivities in dipolar aprotic solvents, as typified by dimethyl sulfoxide (Table 5.4):

$$CN^- > N_3^- > Cl^- > Br^- > I^- > SCN^-$$

The hydration number (the number of water molecules intimately associated with the salt) of the quaternary ammonium salt is very dependent upon the anion. The change in the order of reactivity is thus believed to be due to the hydration of the anion: the highly hydrated chloride and cyanide ions are less reactive than expected, and the poorly hydrated iodide fares better under phase transfer conditions than in homogeneous reactions. Methanol may specifically solvate the anions via hydrogen bonding, and this effect is responsible for the low reactivity of more polar nucleophiles in that solvent.

5.5.3.3   The Role of Water in Phase Transfer Reactions

The optimum concentration of the inorganic substrate, and the role of water itself in phase transfer reactions is a complex issue. Even a simple anion displacement

Table 5.4   Relative reactivities (normalized to iodide) for a series of nucleophiles under phase transfer, homogeneous dipolar aprotic, and homogeneous protic conditions, and the hydration number of the quaternary onium−anion ion pair [43]

Nucleophile	Hydration no.	k_{rel} (PTC)	k_{rel} (DMSO)	k_{rel} (MeOH)
N_3^-	3.0	6.8	25.5	1.03
CN^-	5.0	4.2	63.8	0.69
Cl^-	3.4	0.6	6.8	0.15
Br^-	2.1	1.1	4.3	0.38
I^-	1.0	1.0	1.0	1.0
SCN^-	2.0	0.2	0.5	0.29

reaction shows five distinct regions of activity depending on how much water is present in the aqueous phase [49]. When the inorganic phase consists of less than 5 % water by weight (based on the total weight of the water plus the inorganic salt), the rate limiting step is the dissolution of the solid in the aqueous film—a phenomenon termed *thin layer phase transfer catalysis*. No reaction occurs in complete dryness, but reaction commences on addition of as little as 100 ppm of water. Addition of 5–10 % water causes an increase in the degree of hydration and a consequent decrease in the nucleophilicity of the anion in the organic phase, and this causes a decrease in rate. When there is enough water to saturate the organic phase, but insufficient to completely dissolve the inorganic solid, more water can be added without changing the state of either of the two phases. This leads to the constant rate seen from about 10 to 25 % water content. Once the solid has completely dissolved, further addition of water (25 to 60 %) will increase the hydration of the ions in the aqueous phase and reduce the reaction rate. When the aqueous phase contains 60 to 90 % water by weight, all of the ions in both phases are completely hydrated and there is no significant change to the reaction rate in this region. The exact percentage values depend upon the reaction system used. It is clear that the amount of water used in these reactions is critical, and that it is not necessary for all of the inorganic substrate to be dissolved (Figure 5.14).

5.5.4   Using Inorganic Reagents in Organic Reactions

One major advantage of PTC is that it allows organic reagents to be replaced by their inorganic counterparts. For example NaOH may be used instead of *t*-BuONa, and H_2O_2 may be utilized in place of *t*-BuOOH. This is of great benefit as the organic reagents may cost up to 10 times more than the inorganic ones [50].

5.5.4.1   *Phase Transfer Under Alkaline Conditions*

Alkali metal hydroxides (e.g. NaOH) are strong bases available at extremely low cost, and so it is advantageous to use them, where possible, in place of organic amines or alkoxide bases such as potassium *t*-butoxide. An illustrative example is the generation of a carbene from chloroform or other haloform, which would otherwise require a strong alkoxide base and anhydrous conditions to avoid hydrolysis of the carbene. PTC provides a convenient method for the *in situ* generation of carbenes, which may be used for the dichlorocyclopropanation of alkenes, shown in Scheme 5.11. The advantages of this method are clear: the carbene is generated from the CCl_3^- anion in the *organic* phase, where there is a low concentration of water, despite the presence of a separate aqueous phase in the same reaction vessel. This means that the reaction of the carbene with the alkene is much more likely than hydrolysis. Sodium hydroxide is used as the base in place of more expensive metal alkoxides because the aqueous phase has

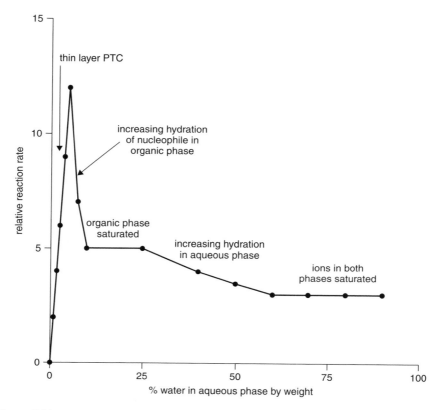

Figure 5.14 The effect of water concentration on the rate of phase transfer reactions

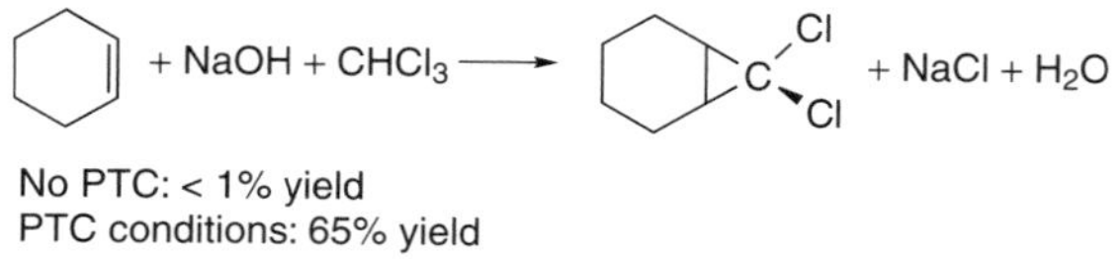

No PTC: < 1% yield
PTC conditions: 65% yield

Scheme 5.11

no detrimental effect on the reaction. The only solvents present are water, and the organic reactants, $CHCl_3$ and the alkene [51].

5.5.4.2 Aromatic Fluorinations Under Phase Transfer Conditions

Many fluorinated aromatic molecules have physiological activity and therefore have agricultural and pharmaceutical applications. An important synthetic pathway to these compounds is the reaction of chloro and nitro aromatics with fluoride salts. Because of the cost of tetraalkylammonium fluorides, the use of alkali metal fluorides is attractive. However, the solubility of inorganic fluoride salts in aprotic solvents is generally very low, and attempted fluorinations in water promote the formation of phenols because the fluoride has a strong polarizing effect on water, and produces hydroxide. PTC might seem an appropriate way of making fluoride available in the reaction system. Many alkylated ammonium and phosphonium salts, however, exhibit a low thermal stability in the presence of strong bases such as fluoride, particularly at the elevated temperatures necessary to effect aromatic nucleophilic fluorine transfer. Tetraphenylphosphonium bromide has a high thermal stability, and has been found to be particularly effective at catalysing the replacement of aromatic chloro, nitro and sulfonyl groups by fluorine (Scheme 5.12) [52].

Rate enhancements for these types of reaction have been reported to be as high as 200-fold, and the selectivity of the reaction was found to be very substrate dependent. These reactions must be conducted in dipolar aprotic solvents in the absence of water. Although tetramethylammonium chloride is too polar to find widespread application as a phase transfer agent, it has good thermal stability, and this, combined with its low cost, has resulted in its large scale industrial use in phase transfer catalysed aromatic nucleophilic fluorinations.

5.5.4.3 Oxyhalogenation of Aromatic Molecules

In a typical halogenation reaction an aromatic substrate reacts with the elemental halogen in the presence of a catalyst or *halogen carrier* such as pyridine, iron(III) bromide or elemental iron itself. These reactions proceed well, but there are certain drawbacks; bromine and chlorine are difficult to handle in the elemental form, and only half of the halogen atoms are incorporated into the product, the rest being lost as highly acidic hydrogen halide waste. The use of onium salts in these reactions as bifunctional phase transfer–weak Lewis acid catalysts has been reported, and this method solves both problems [53]. Aqueous hydrogen peroxide can be used to oxidize the halide back to the halogen, ready for another reaction

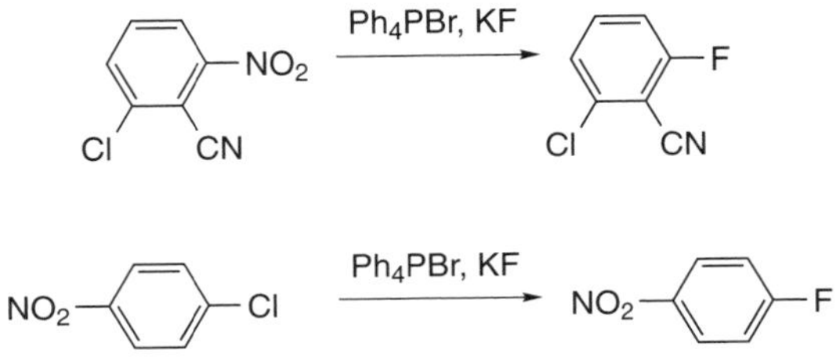

Scheme 5.12

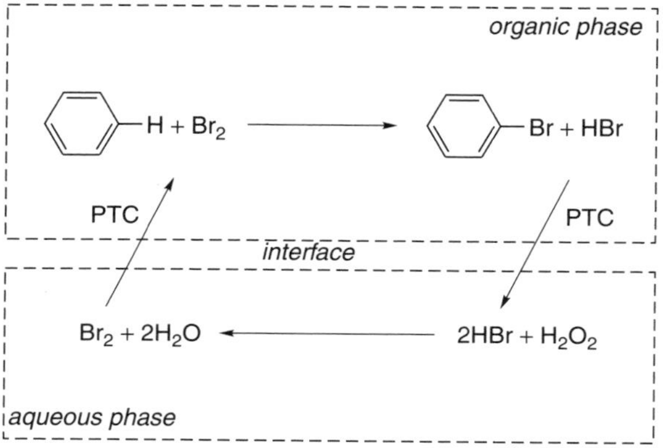

Scheme 5.13

(Scheme 5.13). More conventional Lewis acids such as aluminium chloride and iron(III) chloride catalyse the bromination, but cannot be used in the presence of water. Tetraalkylammonium salts will promote the reaction even in the presence of an aqueous phase. It has been proposed that the phase transfer catalyst can extract both hydrogen peroxide and the hydrogen halide into the organic phase by hydrogen bonding to the onium centre. The handling of elemental halogens may therefore be eliminated, simply by starting with aqueous hydrochloric or hydrobromic acid in place of the chlorine or bromine. This method is advantageous from an environmental viewpoint, as all of the halogen atoms are used and the main by-product is water.

Benzene, toluene, ethylbenzene and chlorobenzene have been shown to be suitable substrates under these conditions, and the reaction rates are 50 to 70 times faster than for uncatalysed reactions. This reaction can be further enhanced by use of methanol as a co-catalyst, which allows bromination of anilines in quantitative yield with complete selectivity for the *para-* isomer [54].

5.6 ORGANOMETALLIC CATALYSIS UNDER PHASE TRANSFER CONDITIONS

As we have seen earlier in this chapter, metal catalysts may be made soluble in water by careful design of the ligands around them. Metal catalysed reactions may also be conducted under phase transfer conditions. Here, by contrast, the metal catalyst usually resides in the organic phase and not the aqueous phase. The use of a phase transfer catalyst in these systems may be to transfer a water-soluble metal catalyst into the organic layer, or else to ensure a supply of water-soluble substrate or reagent to a catalyst already resident in the organic phase. An example of the former is the use of methyltrioctylammonium chloride to extract aqueous $RhCl_3$

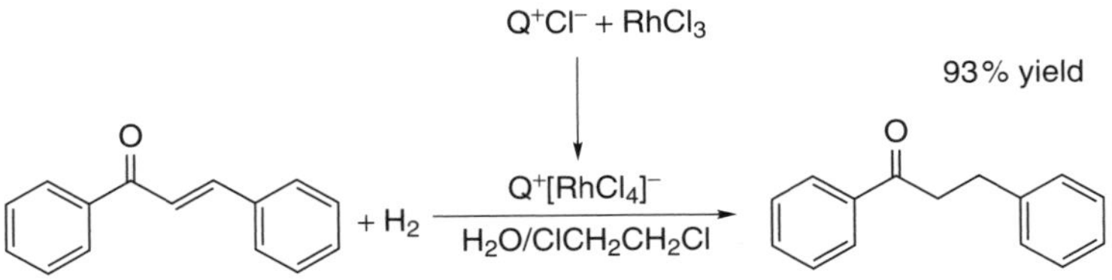

Scheme 5.14

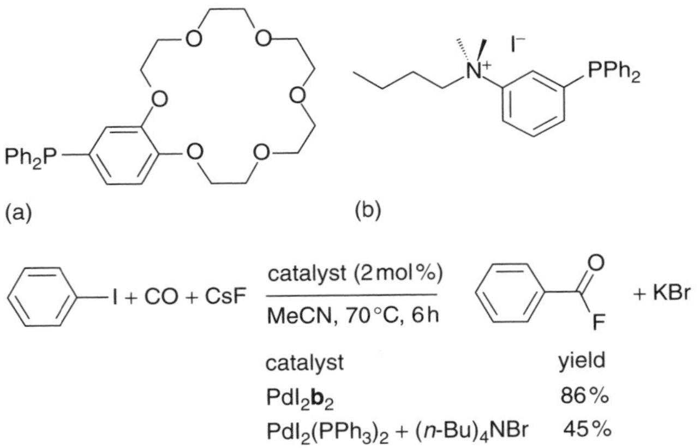

(a)

(b)

catalyst	yield
PdI$_2$**b**$_2$	86%
PdI$_2$(PPh$_3$)$_2$ + (*n*-Bu)$_4$NBr	45%

Scheme 5.15

into 1,2-dichloroethane organic phase by formation of the [RhCl$_4$]$^-$ anion. This is an active hydrogenation catalyst for the selective reduction of a vinyl function whilst leaving the ketone group intact (Scheme 5.14) [55].

In some cases it is advantageous to use a ligand which also acts as a phase transfer agent. Some examples of these are shown in Scheme 5.11. These can lead to accelerated rates for metal catalysed reactions, either because the anion is transferred close to the catalytic centre, or because the PTC group on the ligand brings the catalyst complex closer to the interface or surface. The rates may exceed those observed for a mixture of separate ligand and phase transfer agent [56]. An example of this is the fluorocarbonylation of iodobenzene in the presence of a palladium catalyst, as shown in Scheme 5.15 [57].

5.7 TRIPHASE CATALYSIS

Solid supported reagents offer advantages over liquid reagents because they may be recovered and recycled simply by filtration [58]. This approach has been applied to phase transfer catalysis. If a PTC is immobilized on a solid support then three phases will be present. This is often referred to as *triphase catalysis*, although other systems involving three liquid phases may also be called triphasic [59]. Figure 5.15 shows a schematic representation of the way in which the three reactant phases could come into contact with one another.

Although phase transfer agents have been attached to clays, silica and alumina, the vast majority of studies have used organic polymers, especially polystyrene, as the support. The earliest of these triphase catalysts was prepared from 12 % chloromethylated polystyrene crosslinked with 2 % divinylbenzene by reaction with a tertiary amine. A wide range of triphase catalysts has since been reported, some examples of which are shown in Figure 5.16.

These polystyrene-based catalysts are effective for the cyanide displacements of 1-bromooctane and 1-chlorooctane, and also for the generation of dichlorocarbene from chloroform and aqueous sodium hydroxide, giving quantitative yields of (2,2-dichlorocyclopropyl)benzene from styrene. The catalysts may be recovered simply by filtering the reaction mixture. Unfunctionalized polystyrene does not catalyse these reactions. As well as improving product purification and catalyst recovery, this approach also avoids

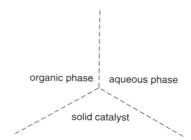

Figure 5.15 Schematic representation of contact between a solid catalyst and two immiscible liquids

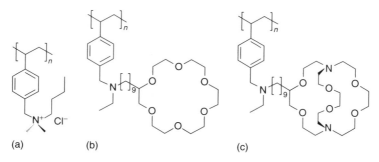

Figure 5.16 Examples of phase transfer catalysts attached to polystyrene, containing (a) a tetraalkylammonium group, (b) a crown ether group and (c) a cryptand group

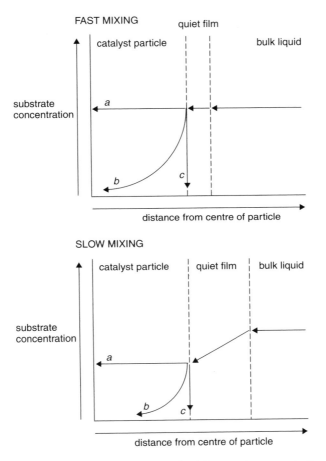

Figure 5.17 Schematic diagram of the effect of mixing on the concentration of substrate in the liquid and solid phases of a triphasic reaction: *a* represents a reaction that is limited only by the intrinsic reactivity; *b* represents a reaction that is limited by a combination of intrinsic reactivity and mass transport effects; *c* represents a reaction which is limited by mass transport only

the risk of producing stable emulsions, which can occur with some soluble PTCs. Polystyrene-bound alkyldimethylammonium salts with longer alkyl chains also catalyse the cyanation of 1-bromooctane [60]. The reaction is first order with respect to the organic substrate and, for polymers with between 1 % and 21 % ring substitution, first order with respect to catalyst. Increasing the degree of polymer ring substitution to very high levels (46 % to 76 %) reduces the activity of the catalysts because the surface of the polymer becomes too polar to be compatible with both aqueous and organic phases. Polystyrene-bound crowns, cryptands, and ammonium and phosphonium salts also effectively catalyse nucleophilic substitution reactions [61]. Like unsupported tetramethylammonium chloride, trimethylammonium functionalized chloromethylated polystyrene is not a good catalyst for these reactions as it is too polar. A longer alkyl chain attached to the quaternary nitrogen centre is normally used to impart some organic character to the catalyst, making it more compatible with the organic phase. These catalysts can be recovered by filtration and reused with little loss in activity.

5.7.1 Mixing Efficiency in Solid–Liquid Reactions

For a triphasic reaction to work, reactants from a solid phase and two immiscible liquid phases must come together. The rates of reactions conducted under triphasic conditions are therefore very sensitive to mass transport effects. Fast mixing reduces the thickness of the thin, slow moving liquid layer at the surface of the solid (known as the *quiet film* or *Nernst layer*), so there is little difference in the concentration between the bulk liquid and the catalyst surface. When the intrinsic reaction rate is so high (or diffusion so slow) that the reaction is mass transport limited, the reaction will occur only at the catalyst surface, and the rate of diffusion into the polymeric matrix becomes irrelevant. Figure 5.17 shows schematic representations of the effect of mixing on the substrate concentration.

Another proposal is that catalysis occurs primarily at the surface. This is not because the rate of reaction is significantly faster than the rate of diffusion, but because the pores of the catalyst become filled with one phase only, which prevents the influx of the other reactant [62].

5.8 CONCLUSIONS

Water is a unique solvent because of its high polarity and ability to form a network of H-bonds. It is immiscible with many organic solvents and is therefore a suitable solvent for use in biphasic reactions in which catalysts are made preferentially soluble in the aqueous phase. Phase transfer catalysis allows the use of aqueous reagents with substrates that have low solubility in water. That water is abundant and totally non-toxic make it the perfect clean solvent, provided that solubility issues can be overcome, and it is in use as a solvent on an industrial scale for polymerization, hydroformylation, and a range of organic chemistry involving PTC. These applications are discussed further in Chapters 7–11.

REFERENCES

1. Matlack A. S. *Introduction to Green Chemistry*, Marcel Dekker, New York, 2001.
2. (a) Chaplin M. F. *Biophys. Chem.* 2000, **83**, 211; (b) Tanaka H. *J. Chem. Phys.* 2000, **112**, 799.

3. (a) Fedyakin N. N. *Kolloid Zh.* 1962, **24**, 497; (b) Derjaguin B. V. and Churaev N. V. *Nature* 1973, **244**, 430; (c) Lippincott E., Stromberg K., Grant W. and Cessac G. *Nature* 1969, **222**, 159; (d) Derjaguin B. V. *Discuss. Faraday Soc.* 1966, **42**, 109.
4. Donohue J. *Science* 1969, **166**, 1000.
5. Vonnegut K. *Cat's Cradle*, Holt, Rinehart & Winston, New York, 1963.
6. (a) Savage P. E. *Chem. Rev.* 1999, **99**, 603; (b) Bröll D., Kaul C., Krämer A., Krammer P., Richter T., Jung M., Vogel H. and Zehner P. *Angew. Chem., Int. Ed. Engl.* 1999, **38**, 2998.
7. Lu J., Brown J. S., Boughner E. C., Liotta C. L. and Eckert C. A. *Ind. Eng. Chem. Res.* 2002, **41**, 2835.
8. Connolly J. *J. Chem. Eng. Data* 1966, **11**, 13.
9. (a) Brown J. S., Gläser R., Liotta C. L. and Eckert C. A. *Chem. Commun.* 2000, **1295**; (b) Chandler K., Deng F., Dillow A. K., Liotta C. L. and Eckert C. A. *AIChE J.* 1998, **44**, 2080.
10. Kalyanam N. and Rao V. G. *Tetrahedron Lett.* 1993, **34**, 1647.
11. Lancaster M. *Green Chemistry: An Introductory Text*, Royal Society of Chemistry, Cambridge, 2002.
12. Bornscheuer U. T. and Kazlauskas R. J. *Hydrolases in Organic Synthesis*, Wiley-VCH, Weinheim, 1999.
13. (a) Aaltonen O. and Rantakylae M. *CHEMTECH* 1991, **21**, 240; (b) Sheldon R. A., Lau R. M., Sorgedrager M. J., van Rantwijk F. and Seddon. K. R. *Green Chem.* 2002, **4**, 147.
14. (a) Fitzpatrick P. A. and Klibanov A. M. *J. Am. Chem. Soc.* 1991, **113**, 3166; (b) Smithrud D. B. and Diderich F. *J. Am. Chem. Soc.* 1990, **112**, 339.
15. Li C.-J. and Chan T.-H. *Tetrahedron Lett.* 1991, **32**, 7017.
16. Cablewski T. Faux A. F. and Strauss C. R. *J. Org. Chem.* 1994, **59**, 3408.
17. Varma R. S. *Green Chem.* 1999, **1**, 43.
18. Hâncu D., Green J. and Beckman E. J. *Acc. Chem. Res.* 2002, **35**, 757.
19. (a) Ahrland S., Chatt J., Davies N. R. and Williams A. A. *J. Chem. Soc.* 1958, 276; (b) Herrmann W. A. and Kohlpaintner C. W. *Inorg. Synth.* 1998, **32**, 8; (c) Paetzold E., Kinting A. and Oehme G. *J. Prakt. Chem.* 1987, **329**, 725; (d) Bianchini C., Frediani P. and Sernau V. *Organometallics* 1995, **14**, 5458; (e) Wan K. and Davies M. E. *Chem. Commun.* 1993, 1262; (f) Schreuder-Goedheijt M., Kramer P. C. J. and van Leeuwen P. W. N. M. *J. Mol. Catal. A* 1998, **134**, 243.
20. (a) Salvesen B. and Bjerrum J. *Acta Chem. Scand.* 1962, **16**, 735; (b) Herd O., Langhans K. P., Stelzer O., Weferling N. and Sheldrick W. S. *Angew. Chem.* 1993, **105**, 1097.
21. Ahrland S., Chatt J., Davies N. R. and Williams A. A. *J. Chem. Soc.* 1958, 276.
22. Herrmann W. A. and Kohlpaintner C. W. *Inorg. Synth.* 1998, **32**, 8.
23. Herrmann W. A., Albanese G. P., Manetsberger R. M., Lappe P. and Bahrmann H. *Angew. Chem.* 1995, **107**, 893.
24. (a) Dreißen-Hölscher B. and Heinene J. *J. Organometal. Chem.* 1998, **570**, 141; (b) Aquino M. A. S. and MacCartney D. H. *Inorg. Chem.* 1988, **27**, 2868; (c) Buhling A., Kramer P. C. J. and van Leeuwen P. W. N. M. *J. Mol. Catal. A* 1995, **98**, 69; (d) Toth I., Hanson B. E. and Davies M. E. *Organometallics* 1990, **9**, 675; (e) Daigle D. J. *Inorg. Synth.* 1998, **32**, 40; (f) Herd O., Heßler A., Hingst M., Tepper M. and Stelzer O. *J. Organometal. Chem.* 1996, **522**, 69; (g) Leik C., Machinitzki P., Nickel T., Schenk S., Tepper M. and Stelzer O. *Z. Naturforsch.* 1999, **54b**, 347; (h) Baxley G. T., Weakley T. J. R., Miller W. K., Lyon D. K. and Tyler D. R. *J. Mol. Catal. A* 1997, **116**, 191; (i) Kant M. and Bischoff S. *Z. Anorg. Allg. Chem.* 1999, **625**, 707; (j) Liliévre S., Mercier F. and Mathey F. *J. Org. Chem.*

1996, **61**, 3531; (k) Stößel P., Mayer H. A. and Auer F. *Eur. J. Inorg. Chem.* 1988, 37; (l) Reetz M. T. *J. Heterocycl. Chem.* 1998, **35**, 1065.

25. (a) Joó F. *Acc. Chem. Res.* 2002, **35**, 738. (b) Karlsson M., Johansson M. and Andersson C. Dalton Trans. 1999, 4187.
26. Moldes I., de la Encarnación E., Ros J., Alverez-Larena A. and Piniella J. F. *J. Organometal. Chem.* 1998, **566**, 165.
27. Allardyce C. S., Dyson P. J., Ellis D. J. and Heath S. L. *Chem. Commun.* 2001, 1396.
28. Martin G. R. and Jain R. K. *Can. Res.* 1994, **54**, 5670.
29. (a) Martin M. T., Prieto I., Camacho L. and Mobius D. *Langmuir* 1996, **12**, 6554; (b) Dargiewicz J., Markarska M. and Radzki S. *Colloids Surf., A* 2002, **208**, 159; (c) Rusin O., Hub M. and Král V. *Mater. Sci. Eng. C* 2001, **18**, 135; (d) Delmarre D., Meallet-Renault R., Bied-Charreton C. and Pasternack R. P. *Anal. Chim. Acta* 1999, **401**, 125; (e) Ford P. C. and Rokicki A. *Adv. Organomet. Chem.* 1988, **28**, 139.
30. (a) Joó F. *Acc. Chem. Res.* 2002, **35**, 738; (b) Hanson B. E. *Coord. Chem. Rev.* 1999, **185**, 795; (c) Roundhill D. M. *Adv. Organomet. Chem.* 1995, **34**, 155; (d) Hermann W. A. and Kohlpaintner C. W. *Angew. Chem., Int. Ed. Engl.* 1993, **32**, 1524.
31. (a) Sasson Y. and Neumann R. (eds) *Handbook of Phase Transfer Catalysis*, Chapman & Hall, London, 1997; (b) Halpern M. E., *Phase Transfer Catalysis: Mechanisms and Syntheses*, ACS Symposium Series 659, American Chemical Society, Washington, DC, 1997; (c) Starks C. M., Liotta C. L. and Halpern M. *Phase Transfer Catalysis: Fundamentals, Applications and Industrial Perspectives*, Chapman & Hall, London, 1994; (d) Makosza M. and Fedoryński M. *Adv. Catal.* 1987, **35**, 375; (e) Rabinovitz M., Cohen Y. and Halpbern M. *Angew. Chem., Int. Ed. Engl.* 1986, **25**, 960; (f) Sasson Y. and Rothenberg G. In *Handbook of Green Chemistry and Technology*, Clark J. H. and Macquarrie D. (eds), Blackwell, Oxford, 2002, p. 206; (g) Tavener S. J. and Clark J. H. *Chem. Ind.* 1997, 22.
32. Ouchi M., Inoue Y., Kanzaki T. and Hakushi T. *Bull. Chem. Soc. Jpn* 1984, **57**, 887.
33. Czech B. P., Pugia M. J. and Bartsch R. A. *Tetrahedron* 1985, **41**, 5439.
34. (a) Starks C. M. *J. Am. Chem. Soc.* 1971, **93**, 195; (b) Brändström A. *Pure Appl. Chem.* 1982, **54**, 1769.
35. Zhu J. and Kayser M. M. *Synth. Commun.* 1994, **24**, 1179.
36. Makosza M. *Pure Appl. Chem.* 1975, **43**, 439.
37. Tundo P. *J. Org. Chem.* 1979, **44**, 2048.
38. Loupy A., Pigeon P., Ramdani M. and Jacquault P. *Synth. Commun.* 1994, **24**, 159.
39. Wovkulich P. M., Shankaran K., Kiegiel K. and Uskovic M. R. *J. Org. Chem.* 1993, **58**, 832.
40. Yang Y. C., Baker J. A. and Ward J. R. *Chem. Rev.*, 1992, **92**, 1729.
41. Landini D., Quinci S. and Rolla F. *Synthesis* 1975, 430.
42. Starks C. M. *J. Am. Chem. Soc.* 1971, **93**, 195.
43. Landini D., Maia A. and Montanari F. *J. Am. Chem. Soc.* 1978, **100**, 2796.
44. Starks C. M. and Liotta C. *Phase Transfer Catalysis: Principles and Techniques*, Academic Press, New York, 1978.
45. Starks C. M., Liotta C. L. and Halpern M. *Phase Transfer Catalysis: Fundamentals, Applications and Industrial Perspectives*, Chapman & Hall, London, 1994.
46. (a) Dehmlow E. V. and Wilkenloh J. *Chem. Ber.* 1990, **123**, 583; (b) Fedoryński M., Ziółkowska W. and Jończyk A. *J. Org. Chem.* 1993, **58**, 6120.
47. (a) Dehmlow E. V. and Dehmlow S. S. *Phase Transfer Catalysis*, 3rd Edition, VCH, Weinheim, 1993; (b) Dehmlow E. V. and Fastabend U. W. E. *J. Chem. Soc., Chem. Commun.* 1993, 124.
48. Mason D., Magdassi S. and Sasson Y. *J. Org. Chem.* 1991, **56**, 7229.
49. Sasson Y., Arrad O., Dermeik S., Zahalka H. A., Weiss M. and Wiener H. *Mol. Cryst. Liq. Cryst. Inc. Nonlin. Opt.* 1988, **161**, 495.

50. Halpern M. *Chemical Market Reporter* 24 May 1999.
51. (a) Makosza M. and Fedoryñski M. *Adv. Catal.* 1987, **35**, 375; (b) Makosza M. *Pure Appl. Chem.* 1975, **43**, 439.
52. Clark J. H. and Macquarrie D. J. *Tetrahedron Lett.* 1987, **28**, 111.
53. Dakka J. and Sasson Y. *J. Chem. Soc., Chem. Commun.* 1987, **19**, 1421.
54. Gervat S., Leonel E., Barraud J. Y. and Ratovelomanana V. *Tetrahedron Lett.* 1993, **34**, 2115.
55. (a) Azran J., Buchman O., Amer I. and Blum J. *J. Mol. Catal.* 1986, **34**, 229; (b) Blum J., Amer I., Zoran A. and Sasson Y. *Tetrahedron Lett.* 1983, **24**, 4139.
56. Amer I. In *Handbook of Phase Transfer Catalysis*, Sasson Y. and Neumann R. (eds), Chapmann & Hall, New York, 1997, and references therein.
57. Okano T., Harada N. and Kiji J. *Chem. Lett.* 1994, 1057.
58. (a) Clark J. H. *Catalysis of Organic Reactions by Supported Inorganic Reagents*, VCH, New York, 1994; (b) Clark J. H., Kybett A. P. and Macquarrie D. J. *Supported Reagents: Preparation, Analysis and Applications*, VCH, New York, 1992.
59. (a) Schmidle C. J. and Mansfield R. C. *Ind. Eng. Chem.* 1952, **44**, 1388; (b) Starks C. M. *J. Am. Chem. Soc.* 1971, **93**, 195; Regen S. L. *J. Am. Chem. Soc.* 1975, **97**, 5956.
60. Regen S. L. *J. Am. Chem. Soc.* 1976, **98**, 6270.
61. Cinouini M., Colonna S., Molinari H., Montanari F. and Tundo P. *J. Chem. Soc., Chem. Commun.* 1976, 394.
62. Hradil J., Svec F., Konák C. and Jurek K. *React. Polym.* 1988, **9**, 81.

6 Supercritical Fluids

Supercritical fluids (SCFs) are best known through their use for the decaffeination of coffee, which employs supercritical carbon dioxide ($scCO_2$). In this chapter, we will demonstrate that SCFs also have many properties that make them interesting and useful reaction media. Firstly, the physical properties of SCFs will be explained, then the specialist equipment needed for carrying out reactions under high temperatures and pressures will be described. Finally, we will discuss issues relevant to the use of SCFs as solvents for reactions.

6.1 INTRODUCTION

Supercritical fluids represent a different type of alternative solvent to the others discussed in this book since they are not in the liquid state. A SCF is defined as a substance above its critical temperature (T_c) and pressure (P_c)[1], but below the pressure required for condensation to a solid, see Figure 6.1 [1]. The last requirement is often omitted since the pressure needed for condensation to occur is usually impractically high. The critical point represents the highest temperature and pressure at which the substance can exist as a vapour and liquid in equilibrium. Hence, in a closed system, as the boiling point curve is ascended, increasing both temperature and pressure, the liquid becomes less dense due to thermal expansion and the gas becomes denser as the pressure rises. The densities of both phases thus converge until they become identical at the critical point. At this point, the two phases become indistinguishable and a SCF is obtained.

Critical points vary widely. Table 6.1 shows a representative sample of critical parameters and it is immediately obvious why carbon dioxide is widely used. With a critical temperature just above room temperature and a critical pressure that is relatively low, the amount of energy needed to render carbon dioxide supercritical is comparatively small. Fluoroform (CHF_3) and difluoromethane also have easily attainable critical parameters, but they are much more expensive than carbon dioxide. Despite its high critical temperature and pressure, supercritical water (scH_2O) is used widely as a destructive medium since it is highly acidic.

---

[1] Although the SI unit of pressure is the Pascal (Pa), it seems to be a convention for chemists working with SCFs to use bar (1 bar = 0.986923 atm = 14.503774 psi = 750.063755 torr = 100 000 Pa).

Chemistry in Alternative Reaction Media   D. Adams, P. Dyson and S. Tavener
© 2004 John Wiley & Sons, Ltd   ISBNs: 0-471-49848-3 (Cloth); 0-471-49849-1 (Paper)

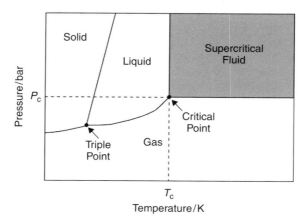

Figure 6.1 A typical phase diagram showing the supercritical region

Table 6.1 Critical data for selected compounds [2]

Substance	$T_c(°C)$	P_c(bar)	ρ_c(g ml^{-1})
CH_4	−82.5	46.4	0.16
CHF_3	26.2	48.5	0.62
CH_2F_2	78.1	57.8	0.42
NH3	132.5	112.8	0.24
CO_2	31.1	73.8	0.47
H_2O	374.2	220.5	0.32
C_2H_6	32.4	48.8	0.20

6.2 PHYSICAL PROPERTIES

Unsurprisingly, a SCF has many physical properties that are intermediate between those of a liquid and a gas. Table 6.2 shows a comparison of typical values for physical properties of a pure substance in different phases.

Table 6.2 Comparison of typical physical properties of gases, SCFs and liquids [3]

Property	Gas	SCF	Liquid
Density(g ml^{-1})	10^{-3}	0.2	1
Viscosity(Pa s)	10^{-5}	10^{-4}	10^{-3}
Diffusivity(cm^2 s^{-1})	0.1	10^{-3}	5×10^{-6}

The density of a SCF is typically less than half that of the liquid state, but two orders of magnitude greater than that of a gas. Viscosity and diffusivity are also temperature and pressure dependent.

Importantly, the properties of the substance can vary within the SCF phase boundaries as the temperature and pressure are varied. For example, the density of carbon dioxide at the critical point (at 74 bar) is in the region of $0.4\,\mathrm{g\,ml}^{-1}$. However, this value increases to a value comparable with that of liquid carbon dioxide as the pressure is increased to 250 bar. Properties may change dramatically with pressure near the critical region. There are generally no discontinuities; the changes start gradually and this gives rise to the so-called, ill-defined *near-critical region*. Hence the density changes sharply but continuously with pressure in this region, but more gradually at higher pressures as shown in Figure 6.2.

Many solvent properties are related to density and vary with pressure in a SCF. These include the dielectric constant (ε_r), the Hildebrand parameter (δ) and π^* [5]. The amount a parameter varies with pressure is different for each substance. So, for example, for scCO$_2$, which is very nonpolar, there is very little variation in the dielectric constant with pressure. However, the dielectric constants of both water and fluoroform vary considerably with pressure (Figure 6.3). This variation leads to the concept of tunable solvent parameters. If a property shows a strong pressure dependence, then it is possible to tune the parameter to that required for a particular process simply by altering the pressure [6]. This may be useful in selectively extracting natural products or even in varying the chemical potential of reactants and catalysts in a reaction to alter the rate or product distributions of the reaction.

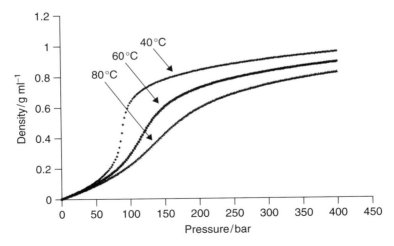

Figure 6.2   The variation in density of CO$_2$ with pressure at 40, 60 and 80 °C (values obtained from [4])

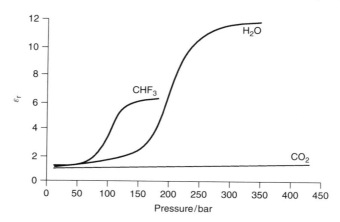

Figure 6.3   Variation in dielectric constant with pressure for scCHF$_3$, scCO$_2$ and scH$_2$O [7]

Supercritical fluids have no surface tension and, like gases, rapidly diffuse to occupy the entire volume of the system. Like a gas, a SCF also mixes perfectly with other gases. As a result, the concentration of gases within a SCF may be much higher than in liquids. The concentration of hydrogen in a supercritical mixture of hydrogen (85 bar) and carbon dioxide (120 bar) at 50 °C is 3.2 M, whereas the concentration of hydrogen in tetrahydrofuran under the same pressure is merely 0.4 M [9]. Supercritical fluids are therefore expected to be good reaction media for hydrogenations and oxidations and this is indeed the case. These reactions are discussed further in Chapters 8 and 9.

6.3   LOCAL DENSITY AUGMENTATION

Supercritical fluids are well behaved when pure (despite their range of properties). However, their behaviour is not always so straightforward when they act as solvents. Within a SCF, short-range inhomogeneities occur around a dissolved solute [11]. This phenomenon is known as to as *local density augmentation* or more simply as *clustering*, although the latter falsely suggests the existence of long-lived clusters. This property is most pronounced near the critical point of the substance due to the large compressibility of the solvent in this region. If these fluctuations in density are of the same order as the wavelength of visible light, *critical opalescence* occurs due to light scattering and the fluid appears cloudy–this has been used as a guide to when a substance becomes supercritical. Hence, it has been found that, in a SCF at a particular bulk density, a probe molecule (e.g. a dye) reports values that would be predicted by theory to occur at higher densities. Thus, the local density felt by the probe is significantly higher than the bulk density [12]. The maximum in this local density augmentation actually occurs at densities below the critical density [13]. Co-solvents, added to

Stereoselectivity in scCO$_2$

The acid-catalysed oxidation of a protected *S*-methyl cysteine, which gives poor diastereoselectivity when oxidized in conventional solvents, shows density-dependent diastereoselectivity as shown in scheme 6.1[8]. Here, *tert*-butyl hydrogen peroxide (TBHP) is used as the oxidant and the reaction is catalysed by an Amberlyst resin (a solid acid). By tuning the pressure at which this reaction was carried out, almost 100 % selectivity to one diastereomer could be achieved (Figure 6.4).

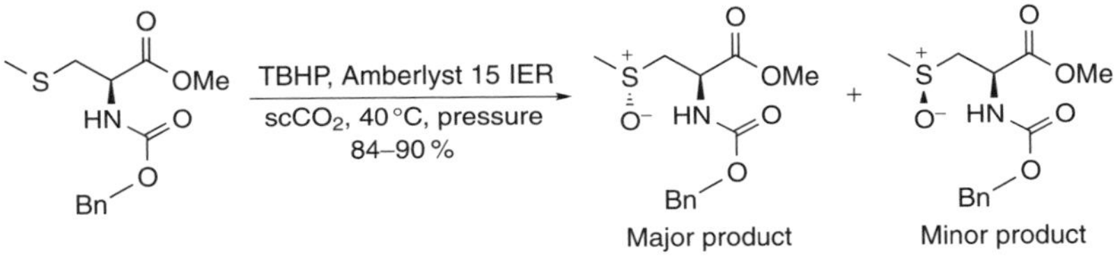

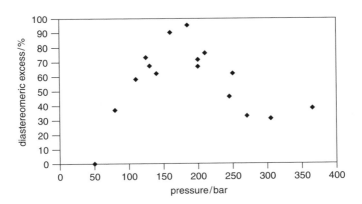

Scheme 6.1

Figure 6.4 shows a scatter plot with x-axis "pressure/bar" and y-axis "diastereomeric excess/%".

Figure 6.4   Effect of pressure on the diastereoselective oxidation of a protected methyl cysteine. Reproduced by Permission of the Royal Society of Chemistry

increase solubility of solutes, also tend to solvate the solutes preferentially, with the result that at lower pressures, where the density is low and there is much more free volume, the local composition can be as much as 10 times the bulk value [14]. However, at high densities, the local composition is not very different from the bulk (Figure 6.5).

These local density augmentations can affect reactions in several ways. The higher local solvent density may change local values of density-dependent properties such as the dielectric constant, thereby influencing reaction rates. It is also

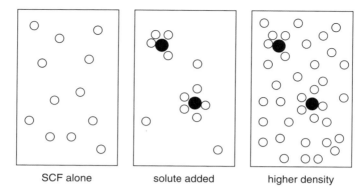

SCF alone	solute added	higher density

Figure 6.5 Solvent clustering in a SCF. At lower densities, the solvent (O) preferentially solvates the solute (•). At higher densities, where there is less free volume, this effect is less pronounced

possible for the cluster of solvent molecules to form a cage around reactants and/or transition state species, altering the progress of some reactions [15].

6.4 SUPERCRITICAL FLUIDS AS REPLACEMENT SOLVENTS

Carbon dioxide is one of the most studied SCF and its use highlights a number of advantages over conventional solvents. Carbon dioxide is nontoxic and, although it is a greenhouse gas, can be obtained in large quantities as a by-product of fermentation and is easily extracted from the atmosphere. From a process perspective, $scCO_2$ and other SCFs offer the advantage that simple depressurization should remove residual SCF and therefore no hazardous solvent effluent is produced. This should lead to the facile separation of products. This absence of solvent residues is of great interest to the pharmaceutical, cosmetics, food and electronic industries where highly pure materials are of paramount importance. Supercritical carbon dioxide is chemically inert under most conditions. It is nonflammable, nonprotic, not strongly Lewis acidic or basic and is inert to radical and oxidizing conditions. Although it can react with some nucleophiles, the general lack of reactivity is very important with regard to the use of $scCO_2$ as a replacement solvent. There are many chemical advantages gained by the use of SCFs as replacement solvents, including high miscibility with gases, and variable solvent parameters such as dielectric constant and density. Local density augmentation may lead to high local concentrations of reagents and the high diffusion rates and altered cage strengths may increase reaction rates.

It should be pointed out that if a reaction or process can currently be performed adequately in a conventional solvent, there is little motivation to switch over to a SCF as the cost could be prohibitive. Instead, financial constraints suggest that

Supercritical fluids allow the formation of species that cannot be made in conventional solvents. For example, η^2-H_2 complexes have been generated by direct reaction of hydrogen with a transition metal carbonyl complex [10]. In order to isolate these compounds, a continuous flow reactor was used and such compounds could be isolated with surprising ease.

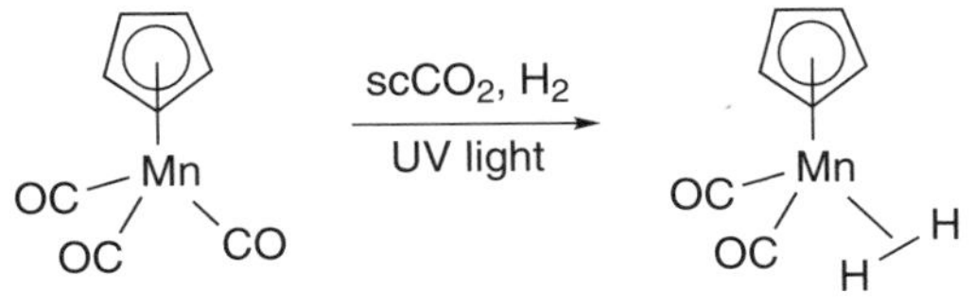

Scheme 6.2

SCFs will find applications in high cost areas such as fine chemical production. Having said that, marketing can also be an issue. For example, whilst decaffeination of coffee with dichloromethane is possible, the use of $scCO_2$ can be said to be natural! Industrial applications of SCFs have been around for a long time. Decaffeination of coffee is perhaps the use that is best known [16], but of course the Born–Haber process for ammonia synthesis operates under supercritical conditions as does low density polyethylene (LDPE) synthesis which is carried out in supercritical ethene [17].

6.5   REACTOR DESIGN

Since reactions in SCFs are carried out at elevated pressures and temperatures, safety is of major importance. Reactor design is very important, not just with regard to being able to carry out the reactions under the required conditions, but also minimizing risks if something were to go wrong. Large-scale use of SCFs requires equipment which is highly specialized and designed especially for the application. However, on a laboratory scale, apparatus tends to share many general features with readily available equipment [18].

There are many aspects to be considered when designing a reactor. These include consideration of the maximum pressure at which the reactor is to be operated, the volume and shape of the cell, which lead to consideration of the method and shape of stirrer required, whether a batch or continuous flow reactor is required, the temperature range of the process including control over the heating rate and removal of any exothermic heat produced during the process, whether any corrosive materials are to be used (scH_2O is capable of corroding steel [19]) and, finally, other issues such as the presence of an optical window for viewing the progress of the reaction or for *in situ* spectroscopic analysis.

Decaffeination of Coffee

One of the best-known examples of the use of $scCO_2$ is for the decaffeination of coffee. This is now an established industrial process (Figure 6.6). In this semi-continuous process, wet green coffee beans are fed into the extractor and then extracted with $scCO_2$. The beans have to be wet to hydrolyse the caffeine before extraction. The caffeine rich CO_2 is then passed through a water wash column, where the caffeine is removed and the CO_2 recycled. The caffeine-rich water is then sent to a reverse osmosis separator where the caffeine is concentrated and recrystallized. For efficiency, the process involves a train of extraction vessels with the freshest $scCO_2$ flowing first through the beans that have been in the system the longest (and are therefore the most extracted). The beans must remain in the extraction vessel for several hours to remove 97 % of the caffeine. Other extractions have also been carried out using $scCO_2$ including essential oils, tannin and nicotine.

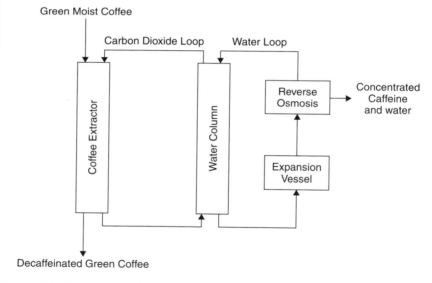

Figure 6.6   Schematic for the decaffeination of coffee on an industrial scale

Typically, reactors are made of stainless steel, which is resistant to most materials although the reactors have been known to catalyse reactions themselves [20]. As mentioned above, scH_2O corrodes steel, and mixtures of $scCO_2$ and water give an acid which is capable of corroding steel relatively quickly under supercritical conditions. Batch reactors generally consist of cylinders of steel with an end cap with a hollowed out section for the reaction cell. Useful design criteria for thick-walled vessels suggest that the cylinder is at least five diameters long, with the wall thickness being at least twice the internal diameter. The lids are generally held in place by bolted flanges and the seal provided by an elastomer or copper o-ring. Despite the fact the elastomer o-rings need frequent replacement, their low cost renders them preferable. These will hold a seal up to approximately 300 bar, deforming under pressure to smooth and fill any irregularities present.

The cell is usually connected to the rest of the system using stainless steel tubing. Both needle and ball valves are available as are check valves that only allow the flow to travel in one direction. The gas is delivered from a cylinder with a pump or compressor in order to provide the required pressures. For safety, two forms of pressure relief are available. Most simply, burst discs can be included to prevent catastrophic system failure due to over pressurization. These are designed

Oxidation in scCO$_2$

When examining the oxidation of alkenes in scCO$_2$, it was found that the stainless steel reactors used were catalysing the reactions [20] In contrast, almost no conversion occurred using toluene as the solvent in standard glassware. Supercritical carbon dioxide was found to be beneficial as a solvent compared to toluene with conversions being much higher in scCO$_2$, probably due to the complete miscibility of oxygen and scCO$_2$ compared to the toluene case, where the amount of oxygen is limited. The postulated mechanism is shown in Scheme 6.3.

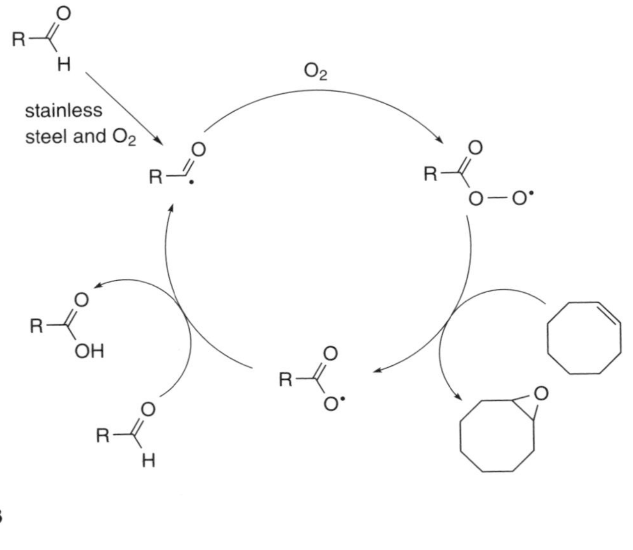

Scheme 6.3

to fail at a known pressure and vent the entire contents of the cell. Secondly, pressure relief valves open and close, releasing excess pressure whilst the system is in use.

Stirring can be achieved simply using a magnetic stirrer bar (steel is not magnetic). If more agitation is needed, perhaps due to a viscous solution or a higher volume cell, packless magnetic drives and stirrers exist. Materials can be

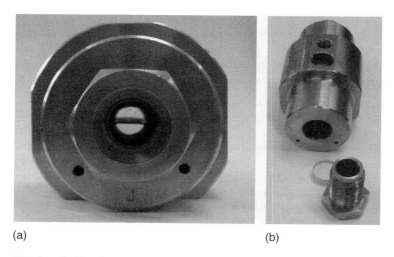

(a)   (b)

Figure 6.7   A typical batch reactor used for small-scale laboratory reactions. The view through the cell is shown (a) together with the dismantled cell (b) Here, the screw thread which holds the window in place can be seen, along with the sapphire window. The holes on top of the cell allow it to be connected to the high-pressure system with the relevant adaptors. (Photograph by Dr A. P. Abbott)

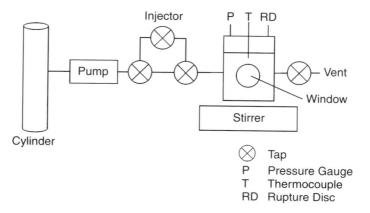

Figure 6.8   A typical set-up for carrying out reactions in scCO$_2$

injected into the reactors via Rheodyne injectors. Should windows be required, these can also be incorporated. Generally, those made of synthetic sapphire will suffice and take the form of a cylinder. These can be screwed in place, the seal again provided by o-rings. Other materials can also be used depending on such issues as the chemical environment to be used and the transmittance of the materials at certain wavelengths. Photographs of a typical reactor are shown in Figure 6.7. A view through the cell is shown in Figure 6.7a demonstrating the use of an optical window together with a view of the cell showing how the window is attached in Figure 6.7b.

A typical set-up for carrying out reactions in scCO$_2$ is shown in Figure 6.8.

6.6 SPECTROSCOPIC ANALYSIS OF SUPERCRITICAL MEDIA

Because of the high pressures and temperatures involved, *in situ* spectroscopic analysis of reactions carried out in supercritical media represents a difficult challenge. However, there are methods available for most of the common techniques [21], with cells being designed specifically for the task at hand.

6.6.1 Vibrational Spectroscopy

Supercritical reaction cells have been adapted for IR analysis by the inclusion of optical windows as described above. The simplest cells consist merely of a small volume cell with a window in two sides, through which an IR beam can be passed [22]. More complex versions also include a mechanism by which the optical pathlength can be varied. The choice of material for the window is obviously important. It must be strong enough to withstand the pressures used and also transmit light at the required frequencies. For example, gallium–gadolinium–garnet windows have been used because of the high modulus of rupture and its wide useable mid-IR range. This material is able to withstand a pressure of 50 MPa. The windows only need to be 5 mm thick to withstand 20.7 MPa, whereas a NaCl window would need to be 10 mm thick [23]. Zinc selenide windows have also been used although the transmission window is narrow (600–700 cm^{-1}) [24]. UV spectroscopy may be carried out in much the same way. For example, cells with quartz windows have been designed to sit in a conventional UV spectrophotometer [25]. Raman spectroscopy can also be carried out using similar cells with quartz windows [26]. A simple alternative is to attach a length of conventional silica capillary tubing (such as that normally used for gas chromatography) to the apparatus [27]. The polyimide coating is removed from a short section and the laser directed onto this piece of the tubing. Good Raman spectra may be recorded because the small diameter allows the scattered light to be efficiently collected. Also, because of the low cost of this tubing, the 'cell' is essentially disposable.

6.6.2 NMR Spectroscopy

High pressure cells for NMR measurements in SCFs have been made out of sapphire [28], fused silica [29] and various plastics [30]. Two main methods exist for carrying out NMR experiments on samples under supercritical conditions. Firstly, a tube can be constructed which is basically analogous to a normal NMR tube (although 10 mm tubes are used rather than 5 mm tubes as for conventional NMR) [31]. Here, a tube of a suitable material (e.g. sapphire) is attached by some means (generally a strong adhesive) to a cap (of a nonmagnetic material such as titanium), which contains adaptors to connect it to high pressure equipment. An example of apparatus for NMR of supercritical reactions is shown in Figure 6.9.

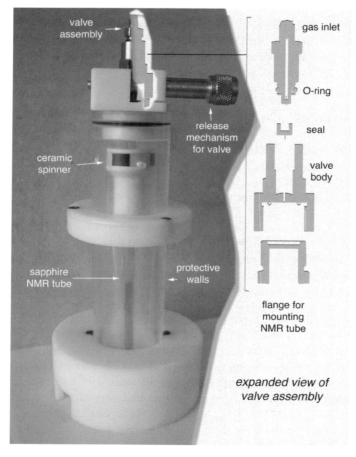

Figure 6.9 An example of a sapphire NMR tube suitable for high pressure NMR Photograph by Gábor Laurenczy

Secondly, a fused capillary tube can be bent repeatedly to permit up to 40 passes in an existing NMR tube. Such tubing, with an internal diameter of about $200\,\mu m$, is coated in polyimide and, where this is removed by the bending process, it is strengthened with cyanoacrylate. The ends of the tubing are glued to ferrules and connected directly to the high-pressure equipment. This equipment fits directly into a normal NMR tube and, if required, can be surrounded by deuterated solvent to lock the signal [31].

6.7   REACTIONS IN SUPERCRITICAL MEDIA

As mentioned above, the physico-chemical properties of SCFs can be manipulated by relatively small changes in the operating temperature and pressure. In principle, it is theoretically possible to tune the solubilities, mass transfer, solvent strength and reaction kinetics of many reacting species present in the SCF. A range of supercritical media are available and the choice of which one is most suitable for the process at hand may come down, as with conventional solvents, to that with the most useful characteristics. Solubility of reagents and catalysts is obviously an issue, with the added complication that this will depend on the chemical functionality, the nature of the SCF and the operating conditions. Many metal complexes are very insoluble in $scCO_2$ (this is not always a drawback – some insoluble catalysts have been found to be highly effective [32]). In general, from the perspective of polarity, the principle of 'like dissolves like' holds. However, the maximum solubility of a polar compound does not necessarily occur in the most polar SCF, for example $scCO_2$ is a better solvent for benzoic acid than $scCHF_3$, with $scCO_2$ behaving as a more polar solvent than expected due to its quadrupole moment. Solubility general increases rapidly near the critical pressure, but, at higher pressures, the increase in solubility is less pronounced. A variety of methods have been used to increase the solubility of polar molecules in $scCO_2$. Co-solvents such as methanol or a fluorinated alcohol may be added [33]. When more than one reagent is used, less polar reagents can themselves act as modifiers, enhancing the solubility of the more polar reagents, and thus avoiding the need for additional co-solvents [34]. For $scCO_2$, an alternative strategy to aid solubility has been to introduce fluorinated substituents, often identical to those used in fluorous solvents (see Chapter 3), onto a ligand or counter-ion. The reason for the increase in solubility caused by the presence of fluorinated substituents has not been fully explained, although there is evidence to suggest that specific solvent–solute van der Waals interactions do exist between CO_2 and certain fluorinated compounds [35]. It has also been suggested that the CO_2 preferentially clusters near the fluorine atoms in the C–F bond and thus fluorinated side-groups may be able to shield the hydrocarbon units from interactions with the solvent. For example, perfluoroalkyl groups have been directly attached to bis(diphenylphosphine)ethane (dppe) analogues [36] to generate a catalyst which was found to be $scCO_2$-soluble with the concentration measuring $7.5 \times 10^{-5}\,\mathrm{mol\,dm^{-3}}$ at a density of $0.55\,\mathrm{g\,ml^{-1}}$, the non-substituted version

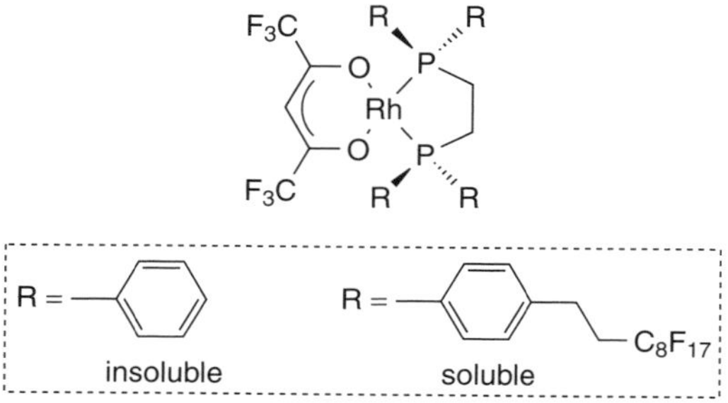

Figure 6.10 The solubility of metal catalysts in scCO$_2$ may be improved by derivatization with perfluoroalkyl groups

being insoluble, (Figure 6.10). Importantly, the electron-withdrawing effect of the perfluoroalkyl groups was sufficiently insulated by the CH$_2$CH$_2$ unit that the reactivity of the metal centre was uncompromised. Perfluoroalkyl substitution has been used successfully to increase the solubility of other metal complexes [37].

Although this approach can be very successful, the high cost of introducing such groups is obviously a disadvantage. However, relatively inexpensive compounds such as polyethers (Figure 6.11) also show enhanced solubility in scCO$_2$ [38]. The carbonyl groups act as Lewis bases and interact with CO$_2$ helping to solubilize the polymer.

The potential for easy separation of the catalyst from products is an attractive feature of reactions in supercritical media because of the ability to tune the solubilizing power of the solvent with density. Recovery of a precious metal catalyst (often including the ligands) is a key issue for the industrial application of such a system. An integrated system can be envisaged whereby the reaction not only occurs in scCO$_2$, but its features are exploited to allow the separation of the products and the recovery of the catalyst in its active form. For example, it is possible to separate products from catalyst after a hydroformylation reaction simply by changing the temperature and pressure. This alters the solubilities of reagents, products and catalysts and so the product aldehydes can be recovered contaminated with less than 1 ppm rhodium. The catalyst remains in the reactor, which can be recharged and used at least five times without loss in activity [39]. As such, SCFs offer an alternative to the biphasic approach to catalyst–product separation techniques [40].

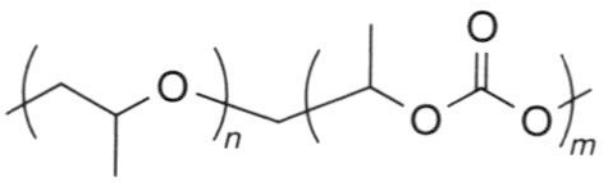

Figure 6.11 Polyether showing high solubility in scCO$_2$

Whilst it is obviously valuable to measure the solubility of reagents in the SCF, it is important to be aware that the solubility in a multicomponent system can be very different from that in the fluid alone. It is also important to note that the addition of reagents and catalysts can have a profound effect on the critical parameters of the mixture. Indeed, at high concentrations of reactants, the mole fraction of CO_2 is necessarily lower and it may not be possible to achieve a supercritical phase at the temperature of interest. Increases in pressure (i.e. further additions of CO_2) could yield a single liquid phase (which would have a much lower compressibility than $scCO_2$). For example, the Diels–Alder reaction (see Chapter 7) between 2-methyl-1,3-butadiene and maleic anydride has been carried out a pressure of 74.5 bar and a temperature of 50 °C, assuming that this would be under supercritical conditions as it would if it were pure CO_2. However, the critical parameters calculated for this system are a pressure of 77.4 bar and a temperature of 123.2 °C, far in excess of those used [41].

Both $scCO_2$ and scH_2O are very attractive solvents for industrial applications because these reaction media are nontoxic, nonflammable and inexpensive. However, $scCO_2$ has the potential to cause problems as a solvent for some reactions. Under reducing conditions, CO may be formed, and the presence of water or alcohol leads to the formation of metal carbonates. Interestingly, CO_2 can interact with catalytically active intermediates and can form a variety of stable coordination compounds. It also readily inserts into metal-hydride, alkyl, alkoxide and amide bonds [42]. A variety of synthetic chemistry has now been reported in $scCO_2$, with the list of reactions constantly growing. Some examples include hydrogenation [43], hydroformylation [44], biotransformations [45], Diels–Alder reactions [46], alkene metathesis [47], Heck and other palladium-mediated reactions [48], rhodium-catalysed hydroboration [49], oxidations and epoxidations [50] and electrophilic aromatic substitution [51]. Supercritical carbon dioxide is also a useful solvent for polymerization reactions [52]. Supercritical water has been used as a reaction media for the production of high-value chemicals by the degradation of compounds of higher molecular weight. For example, lignin has been depolymerized to phenols and other simple ring aromatics [53]. Supercritical water has been used as a solvent for both hydrolysis [54] and dehydration reactions [55], oxidation reactions [56] and palladium-catalysed reactions [57]. Friedel–Crafts reactions have also been carried out in scH_2O supercritical water where the solvent also acts as the catalyst [58].

6.8 CONCLUSIONS

Supercritical fluids have many features that render their use attractive in synthetic chemistry and separations. Their tunable physical properties allow reactions to be carried out under a variety of conditions and, in some cases, the selectivities and rates of reactions may be altered. The list of reactions that have been carried out in SCFs and compared with those in conventional solvents is continually growing.

There are now examples where products and selectivities that cannot be achieved in conventional solvents can be realized by the use of a SCF. However, as noted above, the high running costs which result from carrying out a process at elevated temperatures and pressure may well preclude their use for many reactions.

REFERENCES

1. Jessop P. G. and Leitner W. *Chemical Synthesis using Supercritical Fluids*, Wiley-VCH, Weinheim, 1999.
2. Lide D. R. and Frederiks H. P. R. *CRC Handbook of Chemistry and Physics*, 75th Edition, CRC Press, Boca Raton, 1994.
3. Savage P. E., Gopalan S., Mizan T. I., Martino C. J. and Brock E. E. *AIChE. J.* 1995, **41**, 1723.
4. www.nist.co.uk.
5. (a) Sigman M. E., Lindley S. M. and Leffler J. E. *J. Am. Chem. Soc.* 1985, **107**, 1471; (b) Abbott A. P. and Eardley C. A. *J. Phys. Chem. B* 1999, **103**, 2504; (c) Abbott A. P. and Eardley C. A. *J. Phys. Chem. B* 1998, **102**, 8574; (d) Abbott A. P., Eardley C. A. and Scheirer J. E. *J. Phys. Chem. B* 1999, **103**, 8790.
6. Peck D. G., Mehta A. J. and Johnston K. P. *J. Phys. Chem.* 1989, **93**, 4297.
7. (a) Clifford A. A. *Fundamentals of Supercritical Fluids*, Oxford University Press, Oxford, 1998; (b) Oakes R. S., Clifford A. A. and Rayner C. M. *Perkin Trans.* **1**, 2001, 917; (c) Ed. Jessop P. G. and Leitner W. (eds) *Chemical Synthesis using Supercritical Fluids*, Wiley-VCH, Weinheim, 1999, and references therein.
8. Oakes R. S., Clifford A. A., Bartle K. D., Thornton-Pett M. and Rayner C. M. *Chem. Commun.* 1999, 247.
9. Jessop P. G., Hsiao Y., Ikyaria T. and Noyori R. *J. Am. Chem. Soc.* 1996, **118**, 344.
10. Howdle S. M., Healy M. A. and Poliakoff M. *J. Am. Chem. Soc.* 1990, **112**, 4804.
11. (a) Brennecke J. F. and Chateauneuf J. E. *Chem. Rev.* 1999, **99**, 433; (b) Tucker S. C. and Maddox M. W. *J. Phys. Chem. B* 1998, **102**, 2437.
12. Kim S. and Johnston K. P. *Ind. Eng. Chem. Res.* 1987, **26**, 1206.
13. Carlier C. and Randolph T. W. *AIChE J.* 1993, **39**, 876.
14. Kim S. and Johnston K. P. *AIChE. J.* 1987, **33**, 1603.
15. (a) Johnston K. P. and Haynes C. *AIChE. J.* 1987, **33**, 2017; (b) Ikushima Y., Saito N. and Arai M. *Bull. Chem. Soc. Jpn.* 1991, **64**, 282.
16. Zosel K. *Angew. Chem. Int. Ed. Engl.* 1978, **17**, 702.
17. Hooper C. W. In *Catalytic Ammonia Synthesis*, Jennings J. R. (ed.), Plenum Press, New York, 1991, p. 253.
18. (a) Sherman W. F. and Stadtmuller A. A. *Experimental Techniques in High Pressure Research*, Wiley, London, 1987; (b) Jessop P. G. and Leitner W. *Chemical Synthesis Using Supercritical Fluids*, Wiley,New York, 1999.
19. Kim H., Kurata Y. and Sanadu N. *Proc. Electrochem. Soc.* 1999, **99**, 57.
20. Loeker F. and Leitner W. *Chem. Eur. J.* 2000, **6**, 2011.
21. (a) Poliakoff M. and George M. W. *J. Phys. Org. Chem.* 1998, **11**, 589; (b) Kanakubo M., Umecky T., Liew C. C., Aizawa T., Hatakeda K. and Ikushima Y. *Fluid Phase Equil.* 2002, **194–7**, 859; (c) Kanakubo M., T, Aizawa, Kawakami T., Sato O., Ikushima Y., Hatakeda K. and Saito N. *J. Phys. Chem. B* 2000, **104**, 2749; (d) Fremgen D. E., Smotkin E. S., Gerald II R. E., Klingler R. J. and Rathke J. W. *J. Supercritical Fluids* 2001, **19**, 287.
22. Poliakoff M., Howdle S. M. and Kazarian S. G. *Angew. Chem., Int. Ed. Engl.* 1995, **34**, 1275.
23. Ikushima Y., Sato N., Hatakeda K., Ito S. and Goto T. *Chem. Lett.* 1989, **1707**.

24. Meredith J. C., Johnston K. P., Seminario J. M., Kazarian S. G. and Eckert C. A. *J. Phys. Chem.* 1996, **100**, 10837.
25. Yonker C. R., Frye S. L., Kalkworf D. R. and Smith R. D. *J. Phys. Chem.* 1986, **90**, 3022.
26. Kajimoto O., Futakami M., Kobayashi T. and Yamasaki K. *J. Phys. Chem.* 1988, **92**, 1347.
27. Howdle S. M., Stanley K., Popov V. K. and Bagratashvili V. N. *Appl. Spectrosc.* 1994, **48**, 214.
28. Lim Y., Nugara N. E. and King A. D. *J. Phys. Chem.* 1993, **97**, 8817.
29. (a) Pfund D. M. Zemanian T. S., Linehan J. C., Fulton J. L. and Yonker C. R. *J. Phys. Chem.* 1994, **98**, 11846; (b) Yonker C. R., Wallen S. L. and Linehan J. C. *J. Supercrit. Fluids* 1995, **8**, 250.
30. Lamb D. M., Barbara T. M. and Jonas J. *J. Phys. Chem.* 1986, **90**, 4210.
31. (a) Roe D. C. *J. Magn. Reson.* 1985, **63**, 388; (b) Gaemers S. and Elsevier C. J. *Chem. Soc. Rev.* 1999, **28**, 135.
32. Sellin M. F. and Cole-Hamilton D. J. *J. Chem. Soc. Dalton Trans.* 2000, 1681.
33. Xiao J.-L., Nefkens S. C. A., Jessop P. G., Ikariya T. and Noyori R. *Tetrahedron Lett.* 1996, **37**, 2813.
34. Lemert R. M. and Johnston K. P. *Ind. Eng. Chem. Res.* 1991, **30**, 1222.
35. Dardin A., Cain J. B., DeSimone J. M., Johnson C. S. and Samulski E. T. *Macromolecules* 1997, **30**, 3592.
36. Loeker F. and Leitner W. *Chem. Eur. J.* 2000, **6**, 2011.
37. (a) Banet Osuna A. M., Chen W., Hope E. G., Kemmitt R. D. W., Paige D. R., Stuart A. M., Xiao J. and Xu L. *J. Chem. Soc. Dalton Trans.* 2000, 4055; (b) Francio G. and Leitner W. *Chem. Commun.* 1999, 1663; (c) Francio G., Wittmann K. and Leitner W. *J. Organomet. Chem.* 2001, **621**, 130; (d) Hu Y., Chen W., Banet Osuna A. M., Stuart A. M., Hope E. G. and Xiao J. *Chem. Commun.* 2001, 725; (e) Bonafoux D., Hua Z., Wang B. and Ojima I. *J. Fluorine Chem.* 2001, **112**, 101.
38. (a) Sarbu T., Styranec T. and Beckman E. J. *Nature* 2000, **405**, 165; (b) Sarbu T., Styranec T. J. and Beckman E. J. *Ind. Eng. Chem. Res.* 2000, **39**, 4678.
39. Koch D. and Leitner W. *J. Am. Chem. Soc.* 1998, **120**, 13398.
40. Kainz S., Brinkmann A., Leitner W. and Pfaltz A. *J. Am. Chem. Soc.* 1999, **121**, 6421.
41. Lin B. and Akgerman A. *Ind. Eng. Chem. Res.* 1999, **38**, 4525.
42. (a) Jessop P. G., Ikariya T. and Noyori R. *Chem. Rev.* 1995, **95**, 259; (b) Leitner W. *Coord. Chem. Rev.* 1996, **153**, 257; (c) Gibson D. H. *Chem. Rev.* 1996, **96**, 2063; (d) Mason M. G. and Ibers J. A. *J. Am. Chem. Soc.* 1982, **104**, 5153.
43. (a) Xiao J., Nefkens S. C. A., Jessop P. G., Ikariya T. and Noyori R. *Tetrahedron Lett*, 1996, **37**, 2813; (b) Burk M. J., Feng S., Gross M. F. and Tumas W. *J. Am. Chem. Soc.* 1995, **117**, 8277; (c) Hitzler M. G. and Poliakoff M. *Org. Proc. Res. Dev.* 1998, **2**, 137; (d) Kainz S., Brinkmann A., Leitner W. and Pfaltz A. *J. Am. Chem. Soc.* 1999, **121**, 6421.
44. Rathke J. W., Klinger R. J. and Krause T. R. *Organometallics* 1991, **10**, 1350; (b) Jessop P. G., Ikariya T. and Noyori R. *Organometallics* 1995, **14**, 1510; (c) Palo D. R. and Erkey C. *Ind. Eng. Chem. Res.* 1999, **38**, 2163; (d) Koch D. and Leitner W. *J. Am. Chem. Soc.* 1998, **120**, 13398; (e) Bach I. and Cole-Hamiliton D. J. *Chem. Commun.* 1998, **1463**; (f) Guo Y. and Akgerman A. *J. Supercrit. Fluids* 1999, **15**, 63; (g) Francio G. and Leitner W. *Chem. Commun.* 1999, 1663; (h) Palo D. R. and Erkey C. *Organometallics* 2000, **19**, 81.
45. Mesiano A. J., Beckman E. J. and Russell A. J. *Chem. Rev.* 1999, **99**, 623.
46. (a) Chapuis C., Kucharska A., Rzepecki P. and Jurczak J. *Helv. Chim. Acta* 1998, **81**, 2314; (b) Isaacs N. S. and Keating N. *Chem. Commun.* 1992, 876; (c) Matsuo J., Tsuchiya T., Odashima K. and Kobayashi S. *Chem. Lett.* 2000, 178; (d) Paulaitis M. E. and Alexander G. C. *Pure Appl. Chem.* 1987, **59**, 61.

47. Furstner A., Koch D., Langemann K., Leitner W. and Six C. *Angew. Chem. Int. Ed. Engl.* 1997, **36**, 2466.
48. (a) Shezad N., Oakes R. S., Clifford A. A. and Rayner C. M. *Tetrahedron Lett.* 1999, **40**, 2221; (b) Carroll M. A. and Holmes A. B. *Chem. Commun.* 1998, 1395; (c) Morita D. K., Pesiri D. R., David S. A., Glaze W. H. and Tumas W. *Chem. Commun.* 1998, 1397; (d) Gordon R. S. and Holmes A. B. *Chem. Commun.* 2002, 640.
49. Carter C. A. G., Baker R. T., Nolan S. P. and Tumas W. *Chem. Commun.* 2000, 347.
50. (a) Oakes R. S., Clifford A. A., Bartle K. D., Thornton-Pett M. and Rayner C. M. *Chem. Commun.* 1999, 247; (b) Haas G. R. and Kolis J. W. *Organometallics* 1998, **17**, 4454; (c) Haas G. R. and Kolis J. W. *Tetrahedron Lett.* 1998, **39**, 5923; (d) Pesiri D. R., Morita D. K., Glaze W. and Tumas W. *Chem. Commun.* 1998, 1015.
51. Hitzler M., Smail F. R., Ross S. K. and Poliakoff M. *Chem. Commun.* 1998, 359.
52. (a) Shaffer K. F. and DeSimone J. M. *Trends in Polym. Sci.* 1995, **3**, 146; (b) Canelas D. A. and DeSimone J. M. *Adv. Polym. Sci.* 1997, **133**, 103; (c) Kendall J. L., Canelas D. A., Young J. L. and DeSimone J. M. *Chem. Rev.* 1999, **99**, 543.
53. Funazukuri T., Wakao N. and Smith N. M. *Fuel* 1990, **69**, 349.
54. Broell D., Kaul C., Kraemer A., Krammer P., Richter T., Jung M., Vogel H. and Zehner P. *Angew. Chem., Int. Ed. Engl.* 1999, **38**, 2999.
55. Narayan R. and Antal M. J. *J. Am. Chem. Soc.* 1990, **112**, 1927.
56. Savage P. E. *Chem. Rev.* 1999, **99**, 603.
57. Reardon P., Metts S., Crittendon C., Dougherity P. and Parsons E. J. *Organometallics* 1995, **14**, 3810.
58. Chandler C. K., Deng F., Dillar A. K., Liotta C. L. and Eckert C. A. *Ind. Eng. Chem. Res.* 1997, **36**, 5175.

7 Diels–Alder Reactions in Alternative Media

The Diels–Alder reaction is the name given to the addition of an alkene (a 2π-electron system often called the dienophile) and a 1,4-conjugated diene (a 4π-electron system). This so-called $2 + 4$ addition gives a six-membered ring. Diels–Alder reactions can be used for the construction of complex polycyclic compounds, often with a high degree of regio- and stereocontrol [1]. The reaction is broad in scope; the diene may be open-chain or cyclic and can have many kinds of substituents. However, the alkene must have the *s-cis* conformation. The reaction is stereospecifically *syn-* with respect to both the diene and dienophile and thus the relative configuration of the starting materials is retained in the product. An illustrative example of a Diels–Alder reaction is shown in Scheme 7.1.

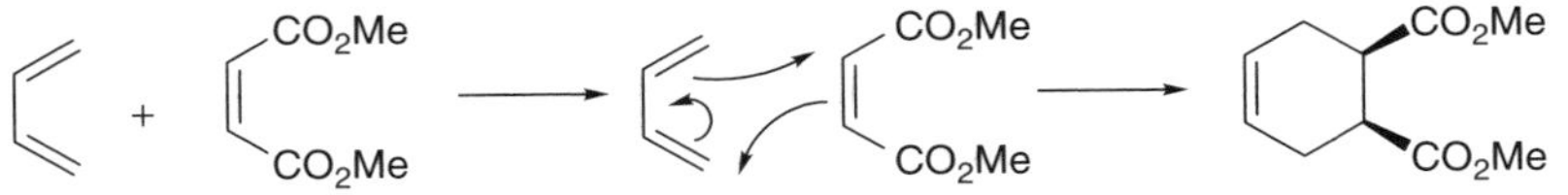

Scheme 7.1

Electron-withdrawing groups attached to the dienophile accelerate the rate of the reaction, donating groups decrease it. For the diene, the opposite is true. Because of the high regiospecificity and the tolerance of a wide number of functional groups, the Diels–Alder reaction is of considerable synthetic importance.

The mechanism of the Diels–Alder reaction involves σ-overlap of the π-orbitals of two unsaturated systems. One molecule must donate electrons, from its highest occupied molecular orbital (HOMO), to the lowest unoccupied molecular orbital (LUMO) of the other. Also, the two interacting orbitals must have identical symmetry i.e. the phases of the terminal p-orbitals of each molecular orbital must match. There are two possible ways for this to happen: the HOMO of the diene combining with the LUMO of the dienophile, and the LUMO of the diene with the HOMO of the dienophile (Figure 7.1).

With the right compounds, the *syn* addition of the diene and dienophile can lead to two possible adducts, depending on whether the larger side of the dienophile ends up under the ring (leading to *endo* addition) or the smaller side is under (giving *exo* addition) as shown in Scheme 7.2. Most of the time, the *endo*

Chemistry in Alternative Reaction Media   D. Adams, P. Dyson and S. Tavener
© 2004 John Wiley & Sons, Ltd   ISBNs: 0-471-49848-3 (Cloth); 0-471-49849-1 (Paper)

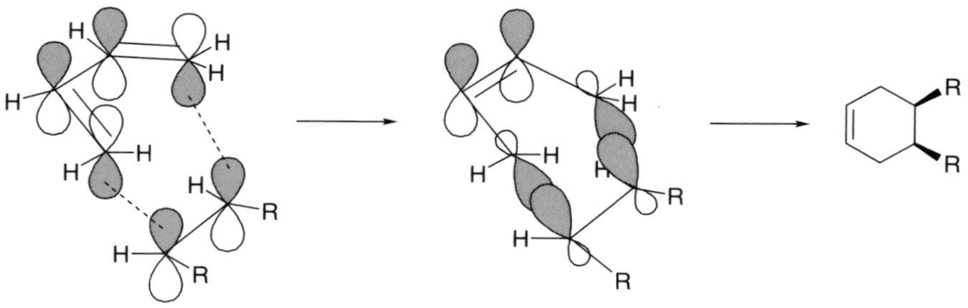

Figure 7.1   Transition state of a Diels–Alder reaction where the LUMO of the diene combines with the HOMO of the dienophile

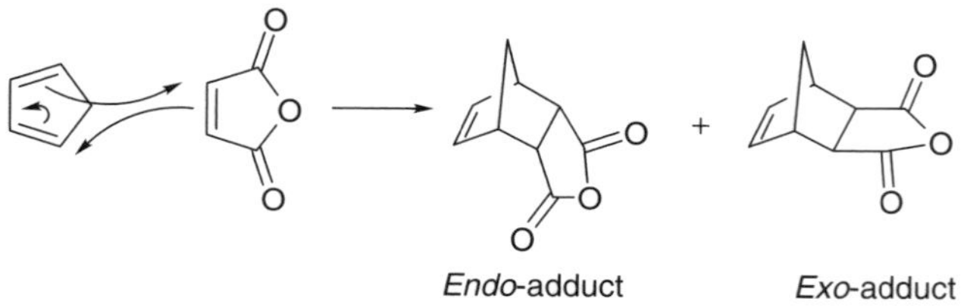

Endo-adduct   Exo-adduct

Scheme 7.2

diastereomer is generally favoured, being the kinetic product of the reaction. This is attributed to the additional orbital overlap between the components in the transition state, which cannot occur in the *exo* transition state.

These reactions do not require solvent, but because high temperatures are often required, a solvent is commonly employed. Since there are no ionic intermediates, the choice of solvent is generally thought to be unimportant and subsequently hydrocarbons are most frequently used. However, there are many occasions where the alternative solvents described in previous chapters can have a dramatic effect on the reaction.

7.1   DIELS–ALDER REACTIONS IN WATER

Despite the fact that Diels and Alder carried out a cycloaddition in water [2], it was not until 1980 that it was reported that large accelerations in the rates of the Diels–Alder reaction could be achieved in water [3]. In addition, selectivity towards the *endo* product was also increased [4]. For example, a 700-fold acceleration in the rate of reaction between cyclopentadiene and 3-buten-2-one (Scheme 7.3) was found in water as compared to reaction in 2,2,4-trimethylpentane. The addition of lithium chloride as a salting-out reagent endo   exo

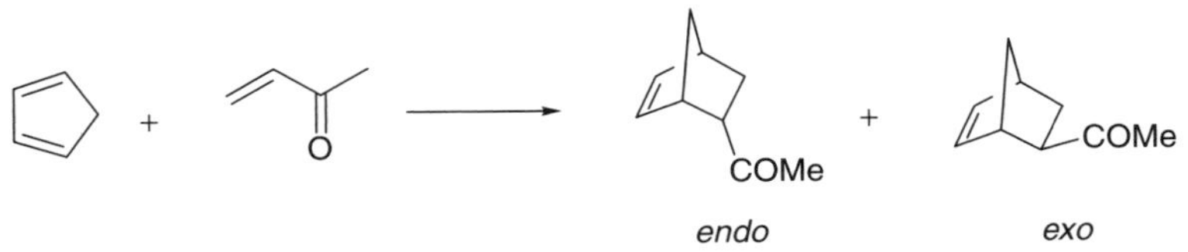

Scheme 7.3

Table 7.1 Rates for reaction between cyclopentadiene and 3-buten-2-one at 40 °C [3]

Solvent	$k_2(\times 10^5\ \mathrm{M^{-1}\ s^{-1}})$
Iso-octane	6
Methanol	76
Water	4400
Water + LiCl	10 800

increased the rate further. Table 7.1 shows some examples of the rates of reaction in a range of solvents.

The reasons for the beneficial results obtained in water are still not completely understood. It is worth emphasizing that this rate enhancement contradicts the accepted mechanism because the polarity of the solvent should have little effect on concerted reactions where there is little difference in polarity between the initial state, the transition state and the final state. The hydrophobic effect (see Chapter 5) is often cited as being responsible. The hydrophobic effect leads to the aggregation of apolar molecules or apolar groups of large molecules in water as a means of minimizing their exposure to the water molecules. Thus, it is argued that this effect 'compacts' the reagents in the transition state compared to the initial state leading to an enhancement in rate. However, it has also been claimed that the hydrophobic effect *cannot* be responsible for the rate enhancements since intramolecular Diels–Alder reactions, where the reactive groups are already closely packed, also show considerable enhancements [5].

Micelle aggregation has also been suggested as a means of increasing the rate of reaction [6], but this is not always the case since such aggregation can sometimes lead to a reduction in rate. The cohesive energy density (CED), a measure of the energy required to create a cavity in the solvent to accommodate the solute, is thought to be an important factor [7]. The CED is related to the intermolecular forces acting in a solvent and the rates of several Diels–Alder reactions have been correlated with the CED of the solvent. Finally, Diels–Alder reactions are known to be catalysed by Lewis acids [8]. Such catalysts also increase the *endo:exo* ratio. The ability of water to *H*-bond to the activating group of the dienophile is thought to be responsible for part of the rate enhancement as well as the increase in *endo:exo* selectivity in much the same way as a Lewis acid catalyst [9].

Despite the obvious benefits outlined above, water cannot always used as a solvent for Diels–Alder reactions. If both reagents are solid, then the reaction will be very slow, if it occurs at all. Also, some substrates possess water-sensitive groups which precludes the use of water as a solvent.

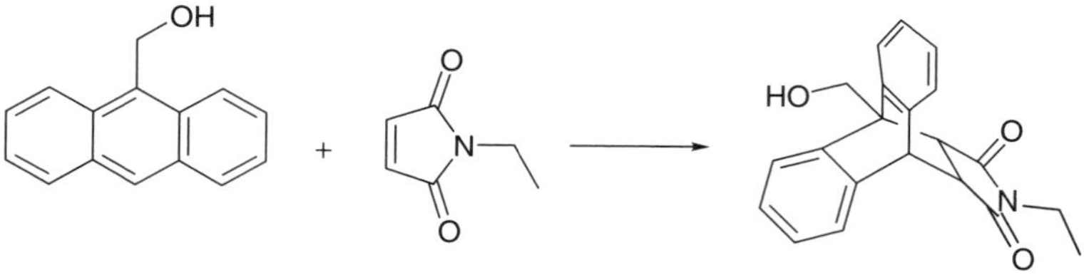

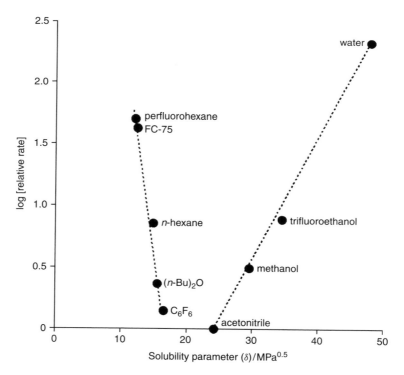

Scheme 7.4

Figure 7.2 The relationship between the solubility parameter (δ), and the relative rates for the reaction shown in Scheme 7.4. (Note that δ for trifluoroethanol did not appear in the original publication [10] and was estimated by correlating π^* for other alcohols)

7.2   DIELS–ALDER REACTIONS IN PERFLUORINATED SOLVENTS

It has recently been shown that the rates of certain Diels–Alder reactions are also accelerated dramatically when carried out in perfluorinated solvents [10]. When the reaction between N-ethylmaleimide and 9-hydroxymethylanthracene (Scheme 7.4) was examined in a range of solvents, significant rate enhancements were found in water and perfluorinated solvents compared to organic solvents.

Interestingly, in fluorinated solvents where the solubility of the reagents is not compromised, for example hexafluorobenzene, the rate of reaction is not increased (Figure 7.2). This implies that there is an inverse relationship between solubility of reagents and the rate of reaction, and that this relationship may be exploited by the use of either water or suitable perfluorinated solvents.

7.3   DIELS–ALDER REACTIONS IN IONIC LIQUIDS

Ionic liquids are excellent solvents for the Diels–Alder reaction providing significant increases in rate and selectivity. Diels–Alder reactions conducted in chloroaluminate ionic liquids show considerable promise. For example, the *endo:exo* ratio for the reaction between cyclopentadiene and methyl acrylate (Scheme 7.5) could be varied by changing the composition of the ionic liquid (see Chapter 4) [11]. Although a high yield was obtained, in order to extract the products it was necessary to quench the ionic liquid in water. This is a considerable disadvantage and for this reason neutral ionic liquids have been the focus of subsequent attention.

It has been shown that Diels–Alder reactions can be carried out successfully in a range of ionic liquids [12]. As highly ordered H-bonding solvents, ionic liquids have the potential for dramatic effects as solvents for such reactions. The range of polarities which can be spanned by varying the cation or anion may be exploited and it has been shown that the *endo:exo* ratio for the reaction between cyclopentadiene and methyl acrylate (Scheme 7.5) is dependent on the polarity of the ionic liquid used [13] (Table 7.2). When the reactions were carried out in a range of ionic liquids, the *endo:exo* values were shown to correlate with the polarity as measured by the E_T^N scale.

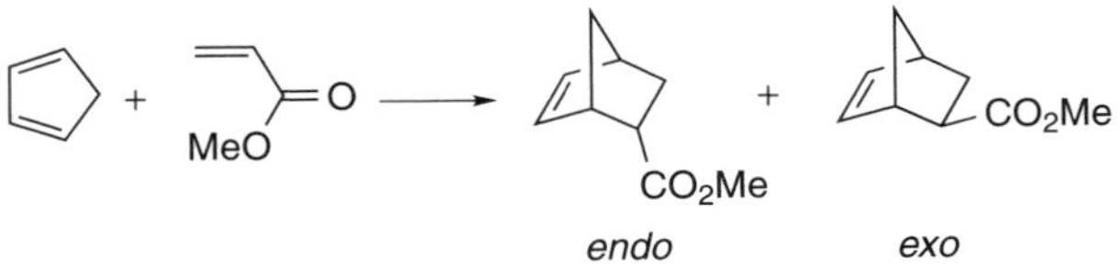

Scheme 7.5

Table 7.2 *Endo:exo* ratios for the reaction between methyl acrylate and cyclopentadiene in a range of ionic liquids [13]

Solvent	E_T^N	endo:exo
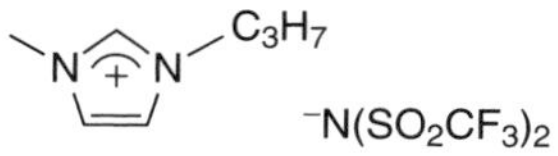 (imidazolium) C_3H_7, $^-N(SO_2CF_3)_2$	0.66	4.1
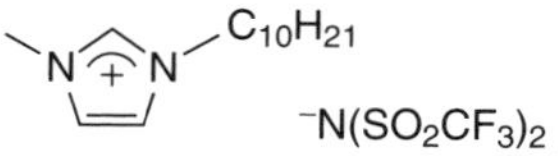 (imidazolium) $C_{10}H_{21}$, $^-N(SO_2CF_3)_2$	0.66	4.3
[bmim][BF_4]	0.67	4.3
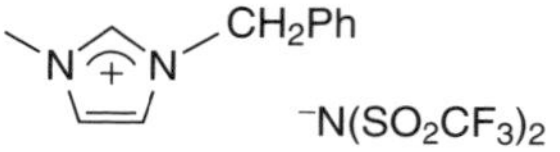 (imidazolium) CH_2Ph, $^-N(SO_2CF_3)_2$	0.67	4.9
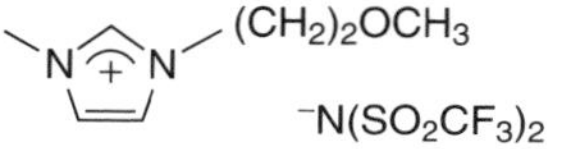 (imidazolium) $(CH_2)_2OCH_3$, $^-N(SO_2CF_3)_2$	0.72	5.7
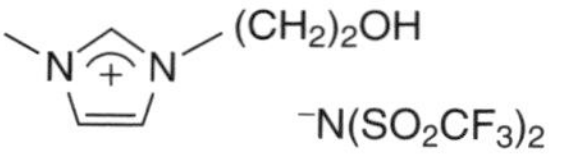 (imidazolium) $(CH_2)_2OH$, $^-N(SO_2CF_3)_2$	0.95	6.1
[Et_3NH]NO_3	0.95	6.7

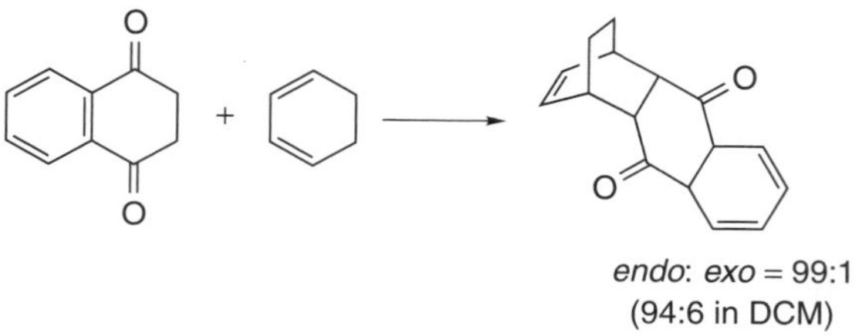

endo: exo = 99:1
(94:6 in DCM)

Scheme 7.6

The advantages of using ionic liquids as solvents for Diels–Alder reactions are exemplified by the scandium triflate catalysed reactions [14] in [bmim][PF_6], [bmim][SbF_6] and [bmim][OTf] for the reaction shown in Scheme 7.6. Whilst the nature of the anion seems to have little effect, all these solvents give rate enhancements for a range of Diels–Alder reactions compared to when the reactions are carried out in dichloromethane (DCM). Also, the selectivity towards the *endo* product is higher than in conventional solvents. As well as the enhanced rates and selectivities, the products can also be removed by extraction with diethyl ether and the ionic liquid and catalyst can immediately be reused. Experiments have shown that reuse of the ionic liquid–catalyst system is possible at least 11 times without loss of activity.

7.4   DIELS–ALDER REACTIONS IN SUPERCRITICAL CARBON DIOXIDE

The Diels–Alder reaction is actually one of the most studied reactions in $scCO_2$. There have been several reports of the effects of pressure on the rate of reaction and product distribution in Diels–Alder reactions carried out in $scCO_2$. There are many examples where Diels–Alder reactions in $scCO_2$ exhibit an increase in reaction rate with increasing pressure [15]. Although the selectivities in $scCO_2$ are often very similar to those achieved in hydrocarbon solvents [16], there have been a number of reports where the use of $scCO_2$ as a solvent can lead to differences in product selectivities. It is potentially possible to control the selectivity of a reaction since the density of the reaction medium can be altered by adjusting the temperature and pressure at which the reaction is carried out. For example, the reaction of methyl acrylate and cyclopentadiene (Scheme 7.5) has been examined in $scCO_2$. Although the effect is small, it is possible to control the selectivity (*endo:exo*) by adjusting the solvent properties. The *endo:exo* ratio was found to change from 2.83 at 100 bar to 2.90 at 300 bar [17]. Similar effects have been found for the same reaction by other researchers [18]. However, in this case, the selectivity was found to drop off again at higher pressures, i.e. to rise up to a maximum and then drop. This has been explained in terms of the adjustment of the position of the nearest-neighbour solvent molecules in the transition state. It has been suggested that the changes in pressure (and hence density) lead to one transition state being favoured over the other [18].

Although the above demonstrated that product control could be achieved in $scCO_2$, the difference in selectivity was relatively small. However, later work using a Lewis acid catalyst, scandium triflate, on the Diels–Alder reaction of *n*-butyl acrylate and cyclopentadiene (Scheme 7.7) showed that the *endo:exo* ratio was again found to rise to a maximum and then decrease again as the pressure, and hence density, was increased (Figure 7.3) [19].

The maximum *endo:exo* ratio occurs above the critical density, suggesting that it is not a critical effect. Again, this effect was rationalized on the basis of a solvent potential tuning phenomenon, where the number of solvent molecules per molecule of substrate is optimized to favour one particular transition state. No

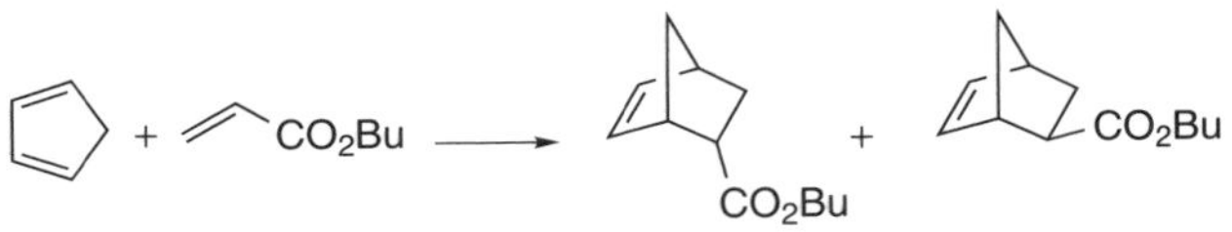

Scheme 7.7

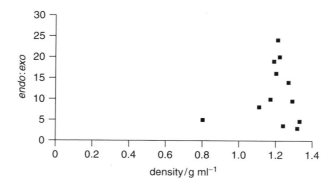

Figure 7.3 Density dependent enhancement of the *endo:exo* ratio for the Sc(OTf)₃ catalysed reaction between *n*-butyl acrylate and cyclopentadiene in scCO₂. Reproduced by permission of the Royal Society of Chemistry

pressure dependent selectivity difference was observed for the scandium triflate catalysed reaction with either the methyl or phenyl acrylate, although the overall selectivity was better than that in toluene.

Other examples of differences in product selectivity have been reported [20], but these were later shown to have been due to poor sampling techniques, highlighting the difficulties in being certain of the phase behaviour in such systems [21].

7.5   CONCLUSIONS

It would appear that water is a remarkable solvent for Diels–Alder reactions giving both rate and selectivity enhancements. There are now many examples of successful reactions being carried out in this solvent. However, water cannot be used for all reactions. Perfluorinated solvents have also been found to give beneficial rate enhancements over organic solvents as have ionic liquids. Interestingly, both ionic liquids and SCFs can be used to tune the selectivities of these reactions, ionic liquids by varying the solvent used and SCFs by altering the density of the solvent.

REFERENCES

1. Nicolaou K. C., Snyder S. A., Montagnon T. and Vassilikoyiannakis G. *Angew. Chem., Int. Ed. Engl.* 2002, **41**, 1668.
2. Diels O. and Alder K. *Ann. Chem.* 1931, **490**, 243.
3. Rideout D. C. and Breslow R. *J. Am. Chem. Soc.* 1980, **102**, 7816.
4. Breslow R. and Maitra U. *Tetrahedron Lett.* 1984, **25**, 1239.
5. Fringuelli F., Piermatti O., Pizzo F. and Vaccaro L. *Eur. J. Org. Chem.* 2001, 439.

6. Grieco P. A., Garner P. and He Z. *Tetrahedron Lett.* 1983, **24**, 1897.
7. Gajewski J. J. *J. Org. Chem.* 1992, **57**, 5500.
8. March J. *Advanced Organic Chemistry*, 4th Edition, Wiley, Chichester, 1992.
9. (a) Lindstrom U. M. *Chem. Rev.* 2002, **102**, 2751; (b) van der Wel G. K., Wijnen J. W. and Engberts J. B. F. N. *J. Org. Chem.* 1996, **61**, 9001.
10. Myers K. E. and Kumar K. *J. Am. Chem. Soc.* 2000, **122**, 12025.
11. Lee C. W. *Tetrahedron Lett.* 1999, **40**, 2461.
12. Fischer T., Sethi A., Welton T. and Woolf J. *Tetrahedron Lett.* 1999, **40**, 793.
13. Dzyuba S. V. and Bartsch R. A. *Tetrahedron Lett.* 2002, **43**, 4657.
14. Sang C. E., Shim W. H., Roh E. J., Lee S.-G. and Choi J. H. *Chem. Commun.* 2001, 1122.
15. (a) Paulaitis M. E. and Alexander G. C. *Pure Appl. Chem.* 1987, **59**, 61; (b) Isaacs N. S. and Keating N. *Chem. Commun.* 1992, 876; (c) Weinstein R. D., Renslo A. R., Danheiser R. L., Harris J. G. and Tester J. W. *J. Phys. Chem.* 1996, **100**, 12337; (d) Ikushima Y., Ito S., Asano T., Yokoyama T., Saito N., Hatakeda K. and Goto T. *J. Chem. Eng. Jpn.* 1990, **23**, 96.
16. Hyatt J. A. *J. Org. Chem.* 1984, **49**, 5097.
17. Kim S. and Johnston K. P. *Chem. Eng. Commun.* 1988, **63**, 49.
18. (a) Clifford A. A., Pople K., Gaskill W. J., Bartle K. D. and Rayner C. M. *Chem. Commun.* 1997, 595; (b) Clifford A. A., Pople K., Gaskill W. J., Bartle K. D. and Rayner C. M. *J. Chem. Soc., Faraday Trans.* 1998, **94**, 1451.
19. Oakes R. S., Heppenstall T. J., Shezad N., Clifford A. A. and Rayner C. M. *Chem. Commun.* 1999, 1459.
20. Ikushima Y., Ito S., Asano T., Yokoyama T., Saito N., Hatakeda K. and Goto T. *J. Chem. Eng. Jpn.* 1990, **23**, 96.
21. Renslo A. R., Weinstein R. D., Tester J. W. and Danheiser R. L. *J. Org. Chem.* 1997, **62**, 4530.

8 Hydrogenation and Hydroformylation Reactions in Alternative Solvents

Addition of molecules across unsaturated organic bonds is an extremely important process that includes reactions such as hydrogenation, hydroformylation, oxidation, hydrocyanation, hydrosilylation and many others. These reactions are often most effectively catalysed by homogeneous catalysts and in this chapter we will focus on hydrogenation (addition of H_2) and hydroformylation (addition of H_2 and CO), which are shown generically in Scheme 8.1.

A number of industrially important compounds require hydrogenation and/or hydroformylation steps at some stage of their synthesis, including pharmaceuticals, herbicides, flavours and fragrances. This chapter commences with an introduction to these reactions, briefly summarizing the mechanisms in which a catalyst brings about the reaction. This is followed by a review of hydrogenation and hydroformylation reactions that take place in alternative solvents, generally under biphasic conditions.

Scheme 8.1

8.1 INTRODUCTION

There are many different homogeneous catalysts available that can be used for hydrogenation and hydroformylation reactions. The most widely used metals that are present in these catalysts are the transition metals, especially Co, Ni, Ru, Rh and Ir [1].

Chemistry in Alternative Reaction Media   D. Adams, P. Dyson and S. Tavener
© 2004 John Wiley & Sons, Ltd   ISBNs: 0-471-49848-3 (Cloth); 0-471-49849-1 (Paper)

Perhaps the best known homogeneous hydrogenation catalyst is *Wilkinson's catalyst*, Rh(PPh₃)₃Cl, named after the Nobel Laureate who discovered this extremely important compound. The mechanism by which Rh(PPh₃)₃Cl catalyses the hydrogenation reaction has been intensively studied and involves a series of steps which are illustrated in the catalytic cycle in Scheme 8.2.

The first step in the catalytic cycle is the dissociation of a phosphine ligand from Wilkinson's catalyst which produces a highly reactive trigonal planar rhodium centre, compound **B**. Oxidative addition of H₂ to **B** affords **C** which then undergoes association of the C=C compound to afford **D**. One of the hydride ligands undergoes transfer to the C=C bond affording a coordinated alkyl as

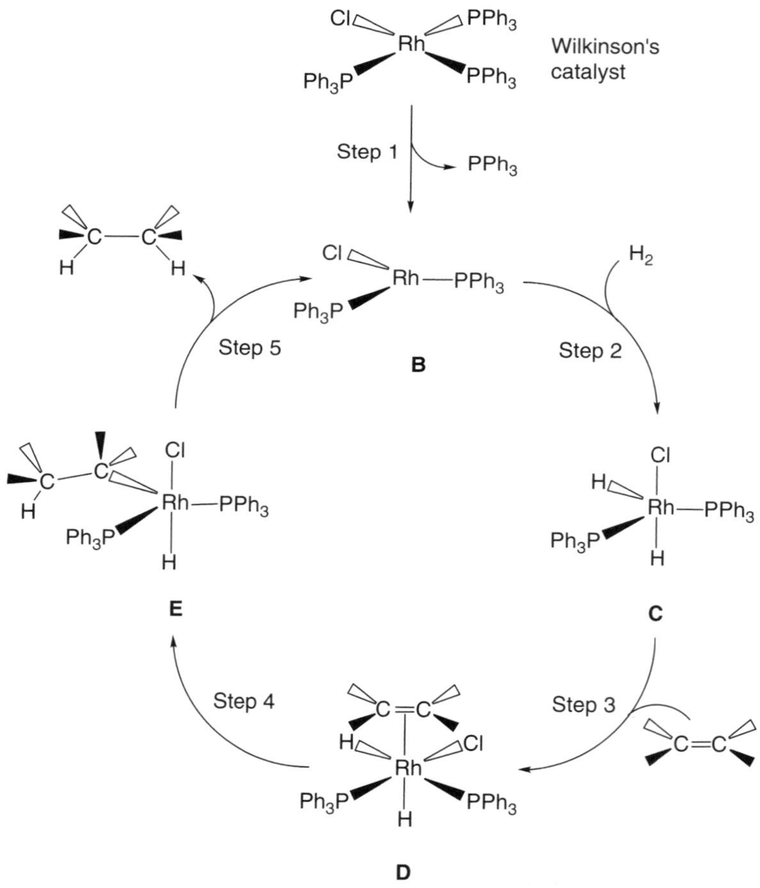

Scheme 8.2 Catalytic cycle for the hydrogenation of C=C bonds using Rh(PPh₃)₃Cl. Step 1, ligand dissociation; step 2, oxidative addition of H₂; step 3, ligand association; step 4, β-hydride transfer; step 5, reductive elimination

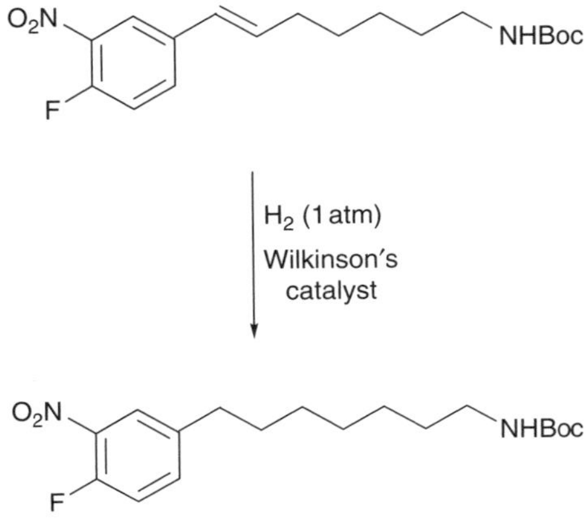

Solvent	Yield (% after 48 h)
Benzene	0
Toluene	20
Methanol	80
Tetrahydrofuran	91
Methanol–tetrahydrofuran	93

Scheme 8.3

shown in **E**. In the final step, reductive elimination of the alkane results in the regeneration of the active catalyst **B** which can participate in another cycle. For a more detailed account of catalytic reaction mechanisms, the reader is referred elsewhere [1].

It is also worth noting that the solvent used can have a considerable influence on the outcome of a hydrogenation reaction. For example, the hydrogenation of the functional alkene shown in Scheme 8.3, using Wilkinson's catalyst, varies considerably according to the conventional organic solvent used [2].

The catalytic cycle for hydroformylation reactions has also been established for certain homogeneous catalysts. Scheme 8.4 illustrates that for $HRh(CO)_2(PPh_3)_2$, although the cycle is the same for the analogous cobalt catalyst.

In the hydroformylation reaction, the fundamental steps that occur are essentially the same, albeit in a different sequence, to those that occur in the hydrogenation reaction, except that there is an additional CO insertion step. In the reaction, a mixture of linear and branched chain products are produced, the resulting composition being defined as the *n:iso* ratio. The *n:iso* ratio obtained is dependent upon the catalyst used and the reaction conditions, and in general, the linear products are of higher value to industry.

8.2 HYDROGENATION OF SIMPLE ALKENES AND ARENES

The hydrogenation of simple alkenes, such as hexene, cyclohexene, cyclohexadiene and benzene, has been extensively studied using biphasic, alternative solvent protocols. These hydrocarbon substrates are more difficult to hydrogenate compared to substrates with electron withdrawing groups. Benzene and alkyl substituted aromatic compounds are considerably more difficult to hydrogenate

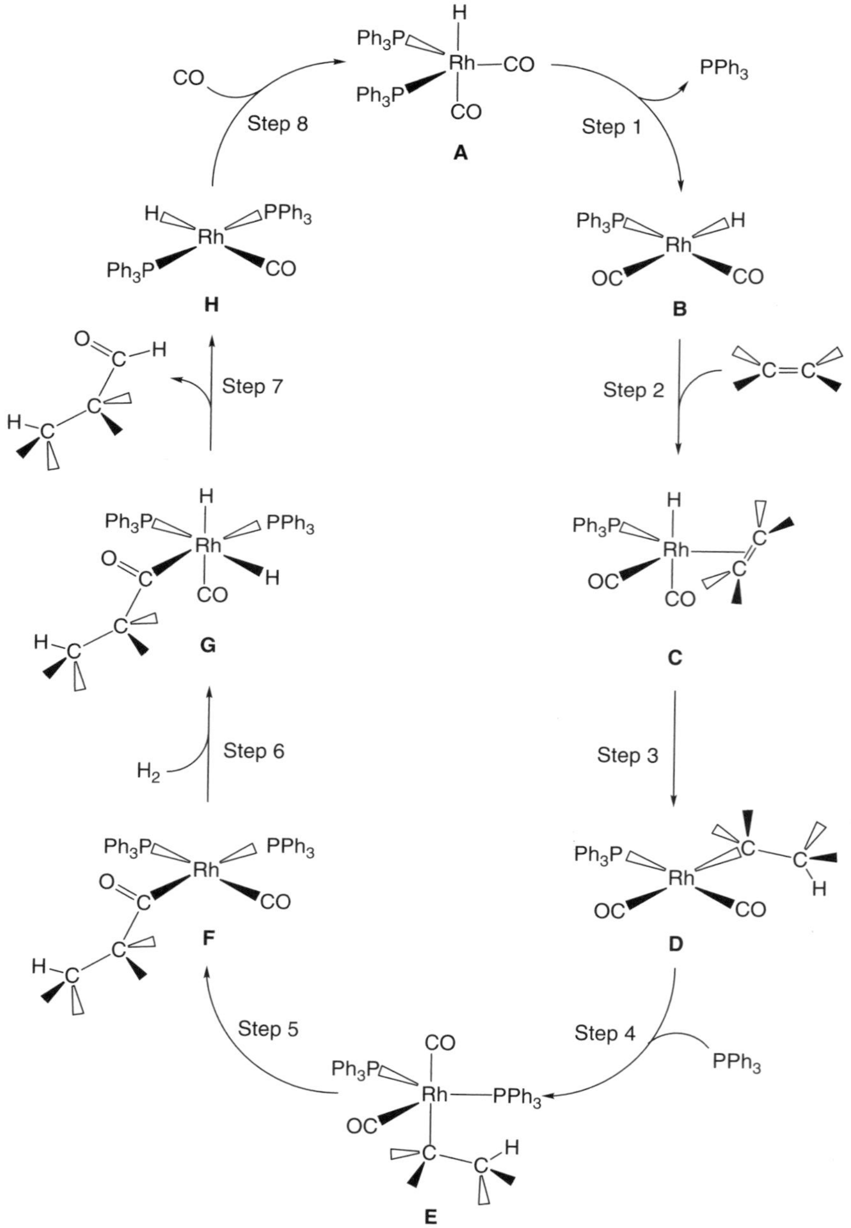

Scheme 8.4 Catalytic cycle for the hydroformylation of C=C bonds using $HRh(CO)_2(PPh_3)_2$.
Step 1, ligand dissociation; step 2, ligand association; step 3, β-hydride transfer; step 4, ligand
dissociation; step 5, CO insertion; step 6, oxidative addition of H_2; step 7, reductive elimination; step 8, ligand association

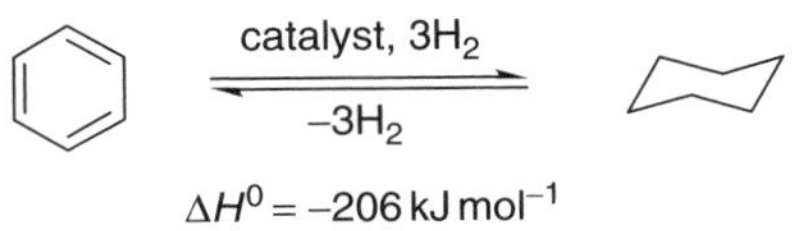

$$\text{catalyst, } 3H_2 \atop -3H_2$$

$$\Delta H^0 = -206 \text{ kJ mol}^{-1}$$

Scheme 8.5

than alkenes, due to the thermodynamics of the hydrogenation reaction (Scheme 8.5) [3].

The hydrogenation of benzene to cyclohexane is an exothermic process with a heat of hydrogenation of $-206\,\text{kJ mol}^{-1}$. The potential intermediates in the hydrogenation, 1,3-cyclohexadiene and cyclohexene, have heats of hydrogenation to cyclohexane of $-230\,\text{kJ mol}^{-1}$ and $-118\,\text{kJ mol}^{-1}$, respectively. The difference in the heat of hydrogenation between benzene and 1,3-cyclohexadiene, namely $24\,\text{kJ mol}^{-1}$, indicates the activation energy required by the hydrogenation reaction, which means the conversion is not possible without a catalyst under ambient conditions. In order to activate the catalyst it is often necessary to heat the system. Unfortunately, the increased temperature shifts the equilibrium of the reaction in favour of the endothermic dehydrogenation reaction. Increased substitution of the aromatic substrate further shifts the equilibrium in favour of the dehydrogenation reaction. To counter this shift in the equilibrium, high pressures of hydrogen are required. High temperatures and pressures only serve to add to the complexity and cost of the hydrogenation process.

8.2.1 Hydrogenation in Water

Hydrogenation reactions in water have been extensively studied and many of the water-solubilizing ligands described in Chapter 5 have been tested in aqueous–organic biphasic hydrogenation reactions. One of the earliest catalysts used was the water-soluble analogue of Wilkinson's catalyst, $RhCl(tppms)_3$ (tppms = monosulfonated triphenylphosphine), but many other catalysts have since been used including $[Rh(cod)(tppts)_2]^+$, $[Rh(cod)_2]^+$ and $[Rh(acac)(CO)_2]^+$ (cod = cyclooctadiene).

Numerous advantages have been observed for hydrogenation reactions carried out using the aqueous-organic biphasic approach. The products have very low solubilities in water and are easily separated. Since both the substrates and products are liquids, they may be used without any additional organic solvent and as such this method fulfils the key goal of avoiding conventional organic solvents altogether.

A reverse set-up has also been developed for the hydrogenation of water soluble substrates. The catalyst is dissolved in the organic phase and the substrates and products in the aqueous phase. Such a protocol has been used to hydrogenate an aqueous solution of 1-butene-1,1-diol, as shown in Scheme 8.6 [4]. Since the

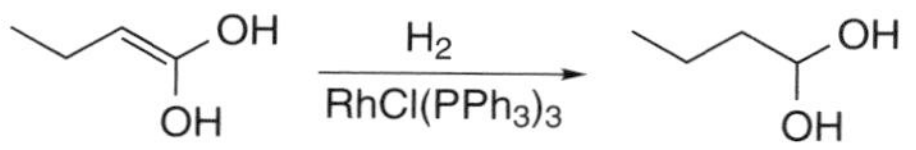

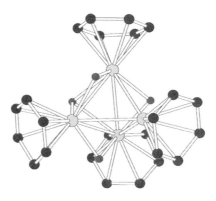

Scheme 8.6

catalyst is immobilized in the organic phase, no modification is required. In this example, Wilkinson's catalyst, $RhCl(PPh_3)_3$, was dissolved in benzene.

This type of reverse set-up has been expanded to catalysts with phosphines containing crown ether substituents (Figure 8.1), with the crown ether acting as a built-in phase-transfer function [5]. Using a catalyst with this phosphine, the hydrogenation of Li^+, Na^+, K^+ and Cs^+ cinnamates in water–benzene solvent mixtures was considerably faster than when the analogous catalyst was used with triphenylphosphine ligands.

The hydrogenation of benzene is a notoriously difficult reaction to catalyse (see above), and biphasic processes are among the most effective known. The tetraruthenium cluster $[H_4Ru_4(C_6H_6)_4]^{2+}$ shown in Figure 8.2 is both water soluble and a highly active catalyst [6]. In addition, water-soluble colloids have also been shown to be extremely effective for the hydrogenation of arenes [7]. For example, a rhodium colloid generated *in situ* hydrogenates benzene and other arenes under aqueous–organic biphasic conditions at room temperature and atmospheric pressure [8].

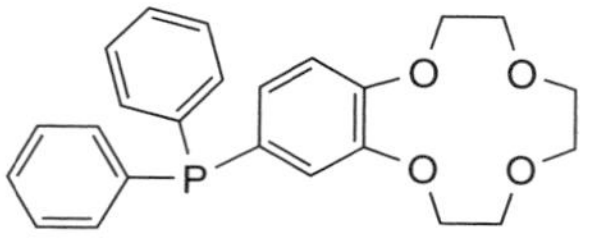

Figure 8.1 A phosphine-substituted crown ether, used as ligand and phase transfer reagent

Figure 8.2 The structure of the benzene hydrogenation catalyst $[H_4Ru_4(C_6H_6)_4]^{2+}$

Hydrogenation reactions in water are not without problems. Water is a good nucleophile and can react with catalysts, particularly highly reactive intermediates in the catalytic cycle, leading to deactivation of the catalyst. In addition, the solubility of hydrogen gas and many organic compounds in water is low compared to conventional organic solvents and other alternative solvents. Table 8.1 lists the solubility of hydrogen in water and some organic solvents. The lower solubility compared to organic solvents means that three to four times higher pressure is needed in order to run a hydrogenation at the same concentration of *dissolved* hydrogen as in the organic solvents under atmospheric pressure. This additional pressure clearly requires greater energy expenditure.

The solubility of some simple terminal alkenes in water is listed in Table 8.2. As the length of the alkyl group increases the solubility of the alkene rapidly decreases. Even with rapid mixing, mass transfer problems due to the low solubility of substrates can occur. As such, alternative solvents to water in biphasic processes are required.

Table 8.1 Solubility of H_2 in water and in organic solvents at 293 K

Solvent	$10^3[H_2](M)$
Water	0.81
Benzene	2.94
Ethanol	2.98
Toluene	3.50
Methanol	3.75

Source: Linke W. F. and Seidell A. *Solubilities of Inorganic and Metal-Organic Compounds*, American Chemical Society, Washington, DC, 1958, Vol. I, p. 1075.

Table 8.2 Solubility of alkenes in water at 298 K

Alkene	Solubility (ppm)
	148
	50
	15
	2.7
	0.6

Source: McAuliffe C. *J. Phys. Chem.* 1966, **70**, 1267.

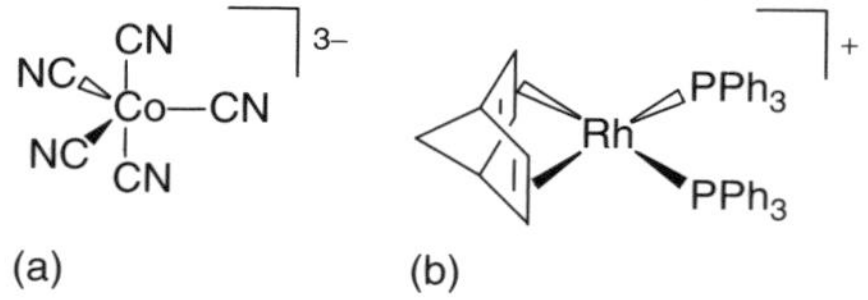

Figure 8.3 The structures of the highly active hydrogenation catalysts (a) $[Co(CN)_5]^{3-}$ and (b) $[Rh(nbd)(PPh_3)_2]^+$

8.2.2 Hydrogenation in Ionic Liquids

Many of the early examples of biphasic reactions using ionic liquids as the immobilization medium employed catalysts that were developed for operation in homogeneous organic processes. For example, Wilkinson's catalyst, $RhCl(PPh_3)_3$, was found to catalyse the hydrogenation of cyclohexene to cyclohexane in [bmim][BF$_4$] [9]. While there are certain advantages to conducting the hydrogenation reaction in this way, such as the higher turnover frequency obtained compared to organic solvents, the neutral catalyst has a relatively high solubility in the organic product phase and is leached during product extraction.

The most effective way to avoid problems of catalyst leaching from an ionic liquid during product extraction is to use a catalyst that is itself ionic. This feature provides high solubility in the ionic liquid and tends to prevent leaching into the organic phase during product extraction. As it happens, there are numerous highly effective hydrogenation catalysts that are salts. The most widely used of these for hydrogenation reactions, that have been extensively studied in other solvents, are $[Co(CN)_5]^{3-}$ and $[Rh(nbd)(PPh_3)_2]^+$ (nbd = norbornadiene), which are shown in Figure 8.3.

The trianionic cobalt catalyst has been successfully employed in the hydrogenation of 1,3-butadiene in [bmim][BF$_4$] [10]. The product from this reaction is 1-butene which is formed with 100 % selectivity. Unfortunately the catalyst undergoes a transformation to an inactive species during the course of the reaction and reuse is not possible. The cationic rhodium catalyst together with related derivatives have been used for numerous reductions, including the hydrogenation of 1,3-cyclohexadiene to cyclohexane in [bmim][SbF$_6$] [11].

In the ideal biphasic hydrogenation process, the substrate will be more soluble or partially soluble in the immobilization solvent and the hydrogenation product will be insoluble as this facilitates both reaction and product separation. Mixing problems are sometimes encountered with biphasic processes and much work has been conducted to elucidate exactly where catalysis takes place (see Chapter 2). Clearly, if the substrates are soluble in the catalyst support phase, then mixing is not an issue. The hydrogenation of benzene to cyclohexane in tetrafluoroborate ionic liquids exploits the differing solubilities of the substrate and product. The solubility of benzene and cyclohexane has been measured in

Asymmetric Hydrogenation Catalysis

The most important goals in hydrogenation catalysis today include enantiose-lective hydrogenations, for which in part the 2001 Nobel prize was awarded, and regioselective hydrogenations. While these types of reactions have been investigated in some detail in alternative solvents, it is difficult to compare the various alternative solvent systems used as a single substrate has not been examined across the range of solvents. Rhodium complexes with the chiral bis-phosphines MeDuPHOS and EtDuPHOS have been used in ionic liquids (and other alternative solvents) to catalyse enantioselective hydrogenations. The results obtained are comparable to those observed in organic solvents [12].

Chiral ionic liquids are also known and the possibility of enantioselective reactions with nonchiral catalysts in such solvents is of considerable interest. Thus far, however, no enantioselective induction has been reported.

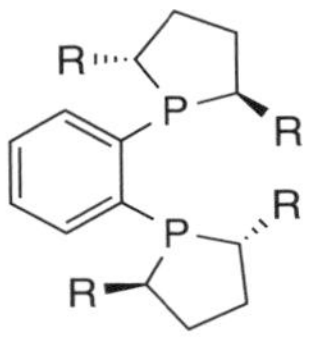

R = Me, MeDuPHOS
R = Et, EtDuPHOS

Figure 8.4 The structure of the chiral bis-phosphine ligands RDuPHOS

four different $[C_x mim][BF_4]$ (C_x = the alkyl chain, i.e. C_4 is a butyl group and the cation is bmim) ionic liquids as shown in Figure 8.5 [13]. From the graph, it may be seen that the solubility of benzene is greater than that of cyclohexane, which is expected, based on their relative polarities (see Chapter 1). In addition, as the length of the alkyl group increases from a methyl group in [mmim][BF_4] to an octyl group in [omim][BF_4], the solubility of benzene rapidly increases. In contrast, the solubility of cyclohexane, which has a lower polarity than benzene, remains effectively constant in all the ionic liquids tested. Since the solubility of benzene is greatest in [omim][BF_4] and the solubility of cyclohexane is almost as low as it is for [mmim][BF_4], then selection of [omim][BF_4] for hydrogenation reactions seems reasonable. In [omim][BF_4], the interface between the benzene substrate and the ionic liquid is greatest allowing optimum mixing of the ben-zene and catalyst without compromising the efficiency of the extraction of the cyclohexane product after reaction.

The dicationic cluster catalyst $[H_4Ru_4(\eta^6\text{-}C_6H_6)][BF_4]_2$, originally developed for biphasic aqueous–organic arene hydrogenation reactions (see Section 8.2.1), has proven to be more effective in ionic liquids presumably due to increased

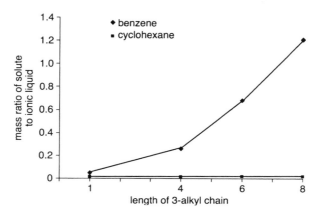

Figure 8.5 The solubility of benzene and cyclohexane in [C$_x$mim][BF$_4$] (C$_x$ = 1, mmim; 4, bmim; 6, hmim; and 8 omim) at 20°C [13]

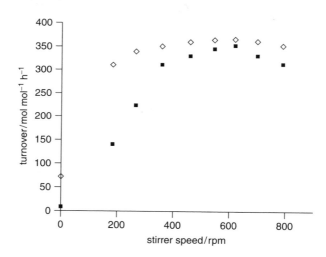

Figure 8.6 The effect of stirrer speed on turnover frequency for the hydrogenation of benzene to cyclohexane. (Note that the turnover reaches its maximum at lower stirrer speed for the ionic liquid). □, ionic liquid; ■, water

solubility of the benzene substrate and hydrogen gas (see Chapter 4) in the ionic liquid compared to water. The effect of the increased solubility of benzene in the ionic liquid has been confirmed by comparing the effect of stirrer speed on turnover in the two solvents. Figure 8.6 shows that the ionic liquid solution is less sensitive to stirrer rate exhibiting close to the maximum turnover at a lower stirrer rate than the analogous reaction in water.

The reactions described above are not truly homogeneous and so their rates are mass-transport limited and depend on the stirrer rate. An ionic liquid–water catalyst system that undergoes a temperature controlled, and reversible, two phase–single phase transition has been developed [14]. At room temperature the ionic liquid 1-octyl-3-methylimidizolium tetrafluoroborate, [omim][BF₄], forms a separate layer to water. On heating to 80 °C, the two phases become completely miscible allowing homogeneous reactions to occur. Using this solvent-system the water soluble substrate butyne-1,4-diol was hydrogenated with [Rh(nbd)(PPh₃)₂] [BF₄] as the catalyst. The reaction was carried out at 80 °C giving a homogeneous single phase solution and on cooling to room temperature, two phases reform, with the ionic liquid phase containing the catalyst and the aqueous phase containing a mixture of 2-butene-1,4-diol and butane-1,4-diol that may be extracted without contamination of the catalyst. The rhodium(I) catalyst was selected for this reaction as it is highly soluble in ionic liquids (since it is a salt), and it is also hydrophobic by virtue of the phosphine ligands.

8.2.3 Hydrogenation in Fluorous Solvents

The hydrogenation of a number of 1-alkenes using fluorous derivatives of Wilkinson's catalyst has been investigated. Initial experiments established the effect of the fluorous ligands on catalytic activity [15]. The ligands evaluated in this study are illustrated in Figure 8.7.

The turnover frequency was found to be greatest for the two fluorous ligands and the control ligand with the trimethylsilyl function was found to be less effective than triphenylphosphine. These fluorous derivatives of Wilkinson's catalyst were used to hydrogenate 1-octene in perfluoromethylcyclohexane (PP2). The reaction was carried out a number of times in order to evaluate the efficiency of the system in terms of recycling and reuse [16]. It was found that as the number

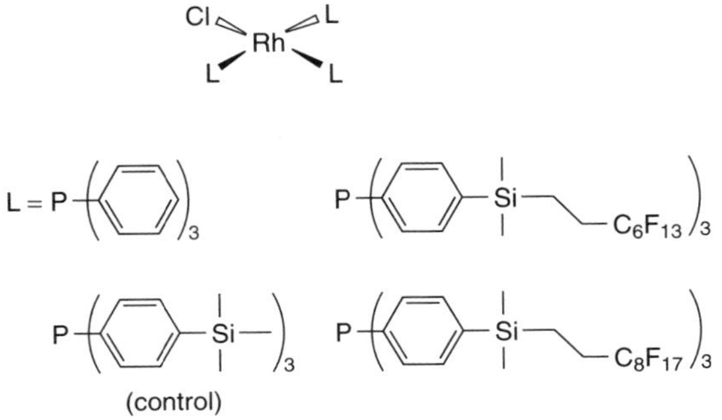

Figure 8.7 Fluorous and control ligands screened to gauge the effect of the fluorous ponytails on catalysis using Wilkinson's catalyst analogues

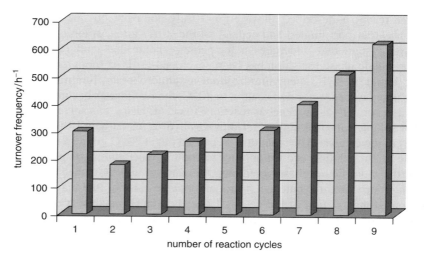

Figure 8.8   The increase in turnover frequency for the hydrogenation of 1-octene using a fluorous derivative of Wilkinson's catalyst as the catalyst (or catalyst precursor)

of reaction cycles increased the turnover frequency of the reaction also increased as shown in Figure 8.8.

The increase in catalytic activity is characteristic of colloidal catalysis. Colloids are known to be highly efficient catalysts, especially for hydrogenation reactions. Under the reducing conditions typified by those containing high concentrations of hydrogen gas, a metal complex can be reduced to form a colloid (or nanoparticle). As the reaction proceeds, or in this case, during each batch, more colloids are generated and hence the rate of the hydrogenation reaction increases. Indeed, colloids were subsequently identified to be the active catalyst using light scattering experiments and electron microscopy [17].

8.2.4   Hydrogenation in Supercritical Fluids

Hydrogenations can be carried out in SCFs using a flow reactor, in which the substrate(s) are passed over a fixed catalyst bed and emerge as products. For example, cyclohexene can be hydrogenated under supercritical conditions in a flow cell using a palladium-based catalyst in either scCO$_2$ or supercritical-propane (see Scheme 8.7) [18]. Performing the reaction in a SCF allows a wide range of reaction parameters to be varied including temperature, pressure and ratio of hydrogen to substrate. Apart from alkenes a wide range of other functional groups have been hydrogenated, including nitriles, alcohols, epoxides, imines, aldehydes and ketones.

Asymmetric hydrogenations have also been carried out in scCO$_2$. α-Enamides have been hydrogenated by using a cationic rhodium complex [19]. Enantiomeric

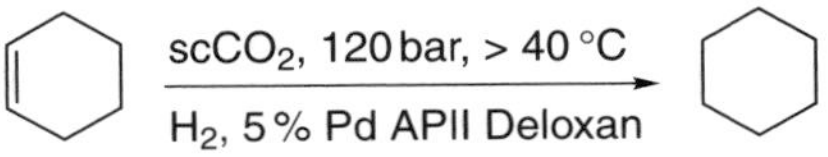

Scheme 8.7

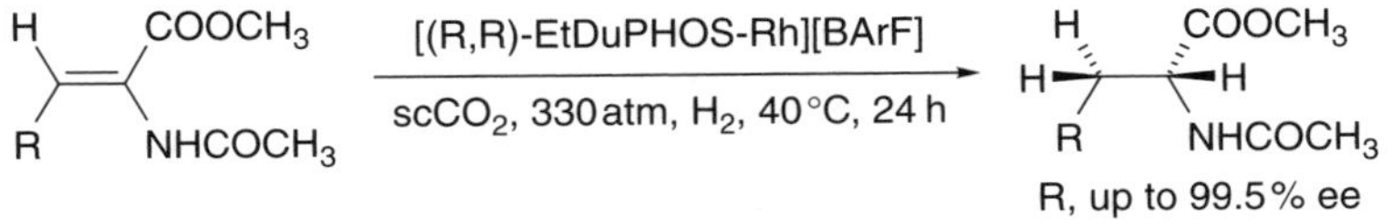

R, up to 99.5% ee

Scheme 8.8

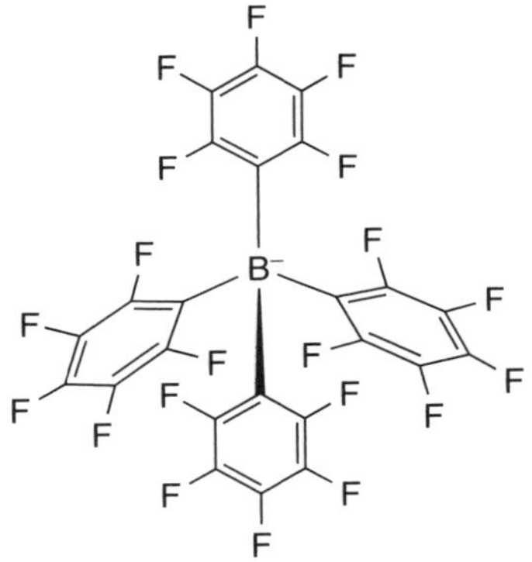

Figure 8.9   BArF

excesses were good to excellent and comparable to those achieved in conventional solvents (see Scheme 8.8).

The cationic rhodium catalyst in Scheme 8.8 is rendered soluble in scCO$_2$ by the anion, BArF, a perfluorinated tetraphenylborate counter-anion (Figure 8.9).

Processes in which catalysts immobilized in ionic liquids have been combined with product extraction using scCO$_2$ show considerable promise [20]. The ionic liquids are poorly soluble in the scCO$_2$ which allows products to be extracted without contamination from the ionic liquid or the catalyst.

8.3   HYDROFORMYLATION REACTIONS IN ALTERNATIVE MEDIA

8.3.1   Hydroformylation in Water

As mentioned in the introduction, hydroformylation is an important industrial process used for the formation of aldehydes from alkenes. Some six million tonnes a year of aldehydes are produced by this route for use in the manufacture of soaps, detergents, plasticizers and pharmaceutical products. The majority of these aldehydes are produced using organic solvents. However, in 1975 a patent was published that reported that various rhodium complexes with sulfonated phosphine ligands catalysed the hydroformylation of propene under aqueous–organic biphasic conditions with virtually complete retention of rhodium in the aqueous phase [21]. This work led to the development of a large scale industrial operation now known as the Ruhrchemie–Rhône Poluenc process for hydroformylation of propene [22]. The research that led to this plant, in a very short time by industrial standards, is described in detail in Chapter 11.

The limitations of hydroformylation reactions in water are the same as those of hydrogenation reactions, i.e. the poor solubility of the substrates (see Section 8.2.1). While aqueous–organic biphasic hydroformylation works well for alkenes with chain lengths up to C_7, the solubility of longer chain alkenes is too low for viable processes. Although simple alkenes are poorly soluble, many functional alkenes have solubilities in water that are sufficiently high to avoid mass transfer problems, but at the same time this can impede separation.

8.3.2 Hydroformylation in Ionic Liquids

Chlorostannate ionic liquids have been used in hydroformylation reactions [23]. Acidic [bmim]Cl-SnCl$_2$ and [1-butyl-4-methylpyridinium]Cl-SnCl$_2$ were prepared from mixing the respective [cation]$^+$ Cl$^-$ with tin(II)chloride in a ratio of 100 : 104, much in the same way that the chloroaluminates are made (see Chapter 4). Both these chlorostannate ionic liquids melt below 25 °C. Addition of Pd(PPh$_3$)$_2$Cl$_2$ to these chlorostannate ionic liquids leads to a reaction medium that catalyses the hydroformylation of alkenes such as methyl-3-pentenoate as shown in Scheme 8.9. The ionic liquid–palladium catalyst solution is more effective than the corresponding homogeneous dichloromethane–palladium catalyst solution. The product was readily separated from the ionic liquid by distillation under vacuum. This is an important reaction as it provides a clean route to adipic acid.

Adipic acid is of considerable importance since it is a precursor to nylon and polyester, which are extensively used in many products. Between two and three million tonnes are produced worldwide each year. Currently, its main method of manufacture is a costly, multistep process involving concentrated nitric acid. Nitrous oxide is produced as a by-product in such quantities that they measurably contribute to global warming and ozone depletion [24]. A cleaner alternative to this process is clearly highly desirable.

One widely studied catalyst system for hydroformylation in organic solvents is Rh(CO)$_2$(acac) combined with triphenylphosphine (or other phosphines) in varying ratios. The same catalyst has been shown to catalyse the hydroformylation of 1-pentene to n- and iso-hexanal in [bmim][PF$_6$] as shown in Scheme 8.10 [25].

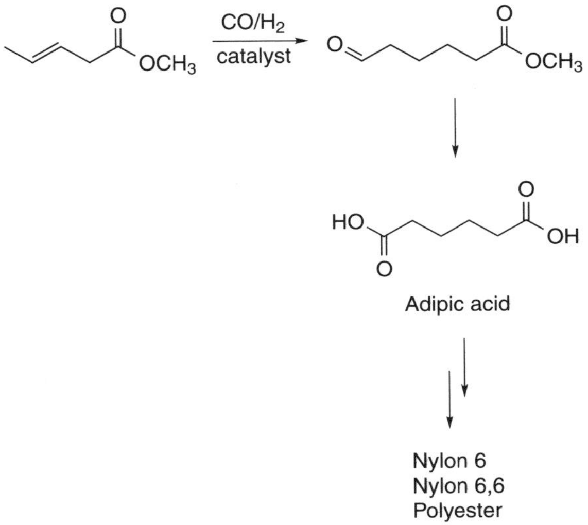

Scheme 8.9

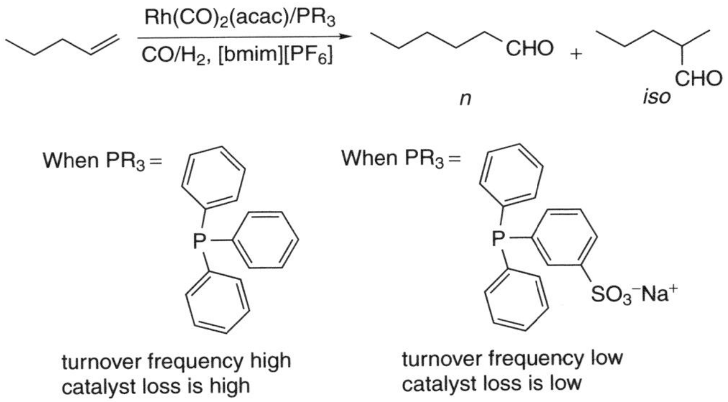

Scheme 8.10

The reaction employing $Rh(CO)_2(acac)$ and triphenylphosphine operates under mild conditions and gives a yield of 99 % and *n:iso* ratio of 3. The main advantage of the reaction is that the product is very easily separated from the ionic liquid as the hexanals are very poorly soluble. The main disadvantage, common when neutral catalysts are employed, is that the catalyst is lost into the organic phase at too high levels to be commercially viable. Catalyst loss can be significantly reduced by replacing the triphenylphosphine ligand with the mono-sulfonated triphenylphosphine ligand (tppms). However, with this phosphine the yield and turnover frequency both decrease significantly.

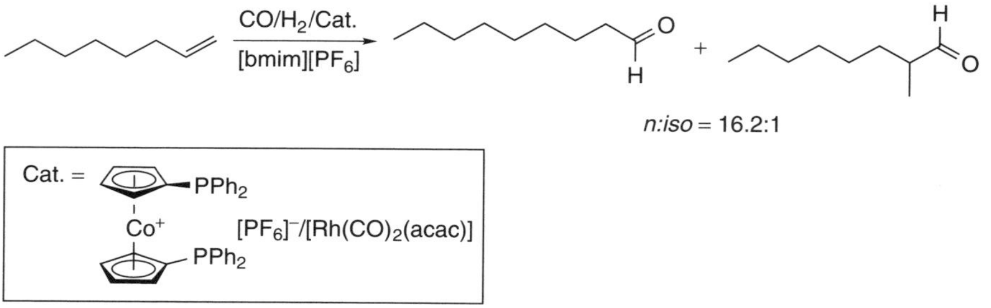

n:iso = 16.2:1

Scheme 8.11

Since the solubility of long-chain alkenes is higher in ionic liquids than in water, there is much interest in finding effective ionic liquid catalysts for the hydroformylation reaction. Bis-phosphines have proved to be particularly useful in hydroformylation reactions and a bis-phosphine with a charged cobaltocenium backbone, analogous to 1,1'-bis(diphenylphosphino)ferrocene (dppf), has been developed specifically for use in ionic liquid hydroformylation reactions [26]. In combination with [Rh(CO)₂(acac)], the 1,1'-bis(diphenylphosphino)cobaltocenium hexafluorophosphate ligand dissolved in [bmim][PF₆] effectively catalyses the hydroformylation of 1-octene as shown in Scheme 8.11.

8.3.3 Hydroformylation in Fluorous Solvents

The fluorous biphasic system is ideally set up for hydroformylation reactions since hydroformylation products tend to be significantly more polar than the starting materials, which facilitates phase separation. The hydroformylation of 1-decene was the subject of the first report on the successful application of fluorous biphase techniques [27]. Here, rhodium complexes of the fluorous derivatized trialkylphosphine, $P(CH_2CH_2C_6F_{13})_3$, were used. This phosphine was chosen on the basis of a molecular modelling study which predicted that two methylene groups would be sufficient to minimize the electron-withdrawing effect of the fluorous ponytail on the electronic properties of the phosphine. Reactions were carried out in a mixture of PP2 and toluene (1 : 1). Catalysts derived from Rh(CO)₂(acac) and each of $P(CH_2CH_2C_6F_{13})_3$, P(alkyl)₃ and triphenylphosphine were all tested under standard reaction conditions in this solvent mixture. Importantly, under the reaction conditions, a single homogeneous phase was formed with all catalyst systems.

The fluorous solvent alone had a minimal effect on the outcome of the reaction. However, with the fluorinated ligand, the n:iso ratio was found to increase with increasing phosphine concentration, reaching a value of almost 8 : 1 at a phosphine to rhodium ratio of 103 : 1. The beauty of this system was demonstrated by its use in a semicontinuous hydroformylation experiment. After each reaction, the mixture was allowed to cool and then the organic layer removed. The fluorous phase was kept in the reactor and reused. Over nine consecutive runs, a total turnover number of greater than 35 000 was measured with only a 4.2 % loss of rhodium. This is equivalent to 1.18 ppm of rhodium lost for each mol of product. This loss was ascribed to the low, but finite, solubility of the catalyst in the product phase. Unfortunately, the free ligand also leached into the organic phase and, since the n:iso was related to the phosphine:rhodium ratio, the n:iso steadily decreased with each run.

Despite this setback, this catalyst system could also be used for the hydroformylation of ethylene and indeed the long-term stability of the catalyst was found to be better than that of the catalyst derived from triphenylphosphine. Hence, this catalyst system allowed the hydroformylation of both high and low molecular weight alkenes under homogeneous conditions combined with facile product separation by simple decantation.

In an effort to overcome the high concentration of the fluorous modified phosphines necessary to achieve high n:iso ratios, the hydroformylation of 1-hexene and 1-octene have been further investigated using a wide range of different phosphine ligands [28]. Since it is well known that triarylphosphine ligands give much better linear selectivity in rhodium based hydroformylation reactions compared to trialkylphosphines [29], fluorous-derivatized triarylphosphines have been prepared and tested. Derivatized triarylphosphites have also been investigated.

Hydroformylation of 1-octene in a toluene-perfluoro-1,3-dimethylcyclohexane biphase with triarylphosphites show much better n:iso ratios and higher initial rates than triphenylphosphine. However, kinetic analysis of the gas uptake suggested that the catalyst was decomposing during the course of the reaction, and rhodium leaching was quite high. Interestingly, observation of the phase behaviour in the system demonstrated that, whilst 1-octene was miscible with the fluorous solvent under the reaction conditions, the product, 1-nonanal, was insoluble. This suggested that the reaction could be carried out in the absence of an organic solvent with the separation benefits still remaining, which would remove the need for a separate distillation to remove the toluene. Under these conditions, reaction rates were found to be significantly improved and the n:iso ratios also increased markedly (7.8 : 1 compared to 4.1 : 1 when toluene was present). The extent of leaching was also greatly reduced, although it was still relatively high.

Investigation of the derivatized triarylphosphine in PP2 demonstrated that this catalyst system was much more stable under the same reaction conditions compared to that derived from the phosphite. The rhodium leaching level was dramatically reduced (0.05 % in one case). Omission of toluene from this system allows development of a process which is nearing the rigorous retention that would be required for commercial application, whilst retaining a high rate and good selectivity to the linear aldehyde product. The refined system also compares well with commercial processes.

The hydroformylation of long chain alkenes using the perfluoroalkyl-derivatized phosphites illustrated in Figure 8.10 has also been examined [30].

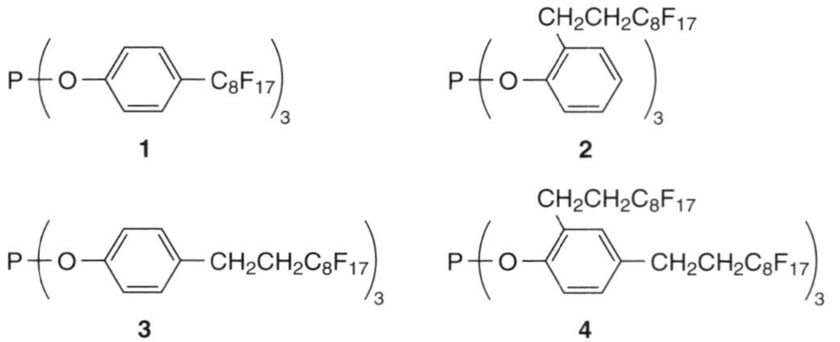

Figure 8.10 Some examples of perfluoroalkyl-derivatized phosphites

Here, CH_2CH_2 spacer groups insulated the aryl rings from the perfluorooctyl groups. This was found to markedly increase the stability of the phosphites towards hydrolysis.

The hydroformylation of 1-decene was investigated using a range of phosphites, organic and fluorous solvents. In general, the activities were found to be comparable to triphenylphosphite catalysed systems. The choice of fluorous solvent was found to be important. For example, the turnover frequencies dropped from 3800 to 2200 when PP2 was replaced by perfluoroperhydrophenanthracene (PFPP). This was attributed to the reduced miscibility of 1-decene with the PFPP compared to with PP2 under the reaction conditions. The *ortho*-functionalized phosphites (**2** and **4** in Figure 8.10) demonstrated different effects compared to the *para*-derivatized compound (**3** in Figure 8.10). The *n:iso* ratio decreased from 3.5 : 1 to 2 : 1 when **2** or **4** were used. The decrease in *n:iso* ratio was thought to be due to a different active species being formed in solution since bulky phosphites are known to form $HRhL(CO)_3$ as the active species [31]. The reuse of the systems was also investigated. The activities were maintained for three cycles (even increasing in some cases). However, the *n:iso* ratio did drop off gradually. This was attributed to gradual leaching of the uncoordinated ligand. The partition coefficients of **2**, **3** and **4** were measured in a $C_8F_{17}H$/1-decene biphase and were calculated to be 19, 19 and 99, respectively[1]. Since all these phosphites do show some solubility in organic solvents, leaching is to be expected. However, decay of the phosphites was also considered as an explanation for the decrease in the *n:iso* ratio. Phosphites are known to be capable of reacting with aldehydes [32], and indeed the phosphites used were not stable in the presence of undecanal.

8.3.4 Hydroformylation in Supercritical Fluids

Although much effort has been directed towards making catalysts soluble in the alternative solvents described in this book, insoluble metal complexes have actu-

---

[1] The partition coefficient P = [concentration in $C_8F_{17}H$]/[concentration in 1-decene].

Asymmetric Hydroformylations in Fluorous Biphase

Hydroformylations have also been carried out in a fluorous biphasic system using a fluorous-derivatized analogue of BINAPHOS [33], one of the best chiral ligands to date for the asymmetric hydroformylation of vinylarenes [34]. The fluorous derivative **5** (Figure 8.11) exhibited comparable catalytic activity, regioselectivity and enantioselectivity to that achieved with BINAPHOS for the hydroformylation of styrene, however, the ligands were preferably soluble in toluene over PP2, rendering their reuse by a fluorous biphasic method impossible. Good solubility was achieved in scCO$_2$ however, although reduced enantioselectivity was observed compared to the nonfluorinated analogue in conventional solvents, which was ascribed to racemization occurring during reaction.

Other fluorinated analogues of BINAPHOS have been used in scCO$_2$. A range of vinyl arenes can be hydroformylated successfully with high enantioselectivity using **6**, comparing well with the analogous reactions in benzene. The homogeneous hydroformylation of vinyl acetate showed improved enantioselectivity compared to the reaction in benzene, the enantioselectivity observed being the highest obtained with this substrate [35].

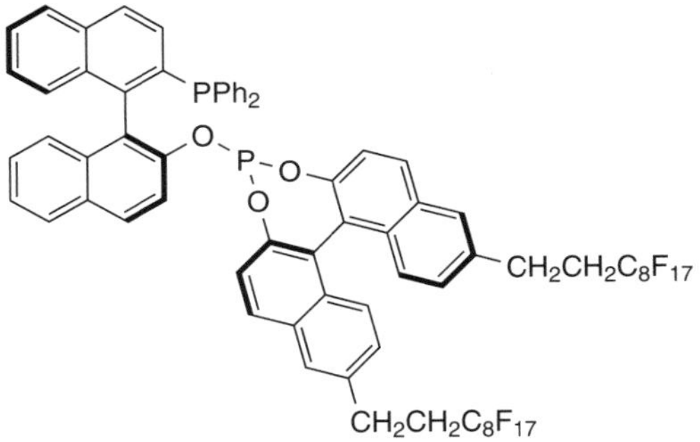

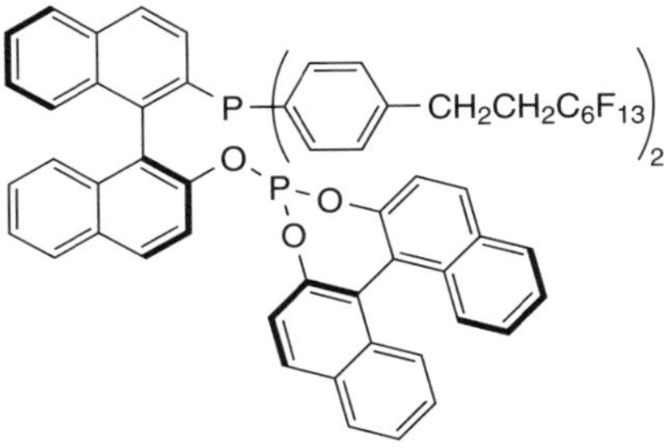

Figure 8.11 Fluorous BINAPHOS ligands

ally been found to be highly efficient catalysts in scCO$_2$. A number of catalysts prepared from [Rh$_2$(OAc)$_4$] were found to be insoluble in scCO$_2$, but when compared to analogous reactions in toluene, the yield is reduced, although the *n:iso* ratio is increased [36]. The insolubility of the catalyst allowed facile separation of the products from the catalyst. After reaction, the stirring was stopped and pure scCO$_2$ used to flush the products into a second reactor where they were collected after depressurization. The first reactor was then recharged with reagents and the reaction repeated. Rhodium was not detected in the products and the catalyst continued to show good activity for several runs.

The hydroformylation of alkyl acrylates (Scheme 8.12) is generally sluggish in conventional solvents. It has recently been shown that a fast and selective hydroformylation of these compounds can be achieved in scCO$_2$ using a catalyst derived from [Rh(acac)(CO)$_2$] and a fluorinated phosphine ligand. In scCO$_2$, the average turnover frequencies for the formation of aldehydes were 10–20 times higher than in toluene. This enhancement was attributed to specific solvent–solute interactions, with the equilibrium of the key unsaturated intermediate being shifted as a result of a carbonyl–CO$_2$ donor–acceptor interaction illustrated in Figure 8.12 [37].

Another way of getting around the problem of the separation of the catalyst from the substrate is via use of a flow reactor [38]. Supercritical carbon dioxide has been used successfully as a medium for the hydroformylation of 1-octene using an immobilized rhodium catalyst. The catalyst is covalently fixed to silica through the modifying ligand *N*-(3-trimethoxysilyl-*n*-propyl)-4,5-bis(diphenylphosphino)phenoxazine (Figure 8.13). Selectivity was found to be

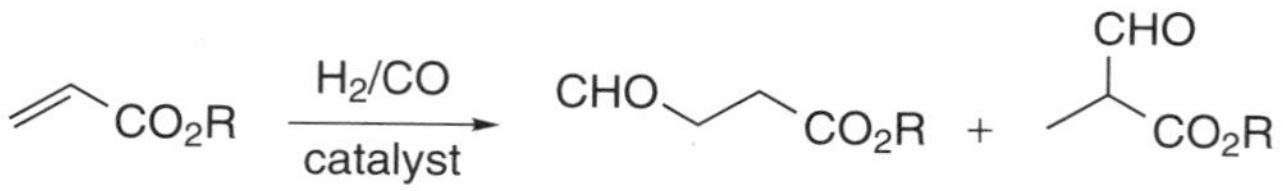

Scheme 8.12

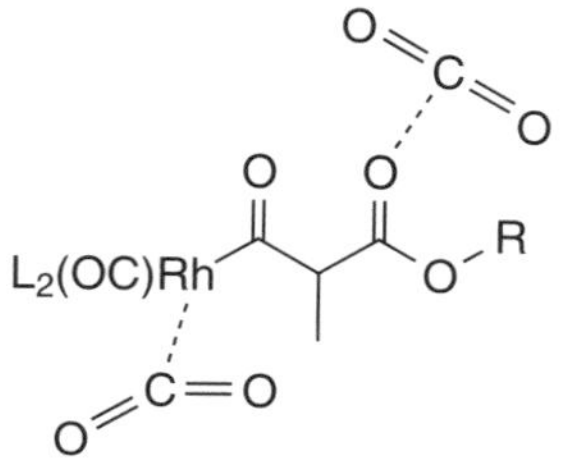

Figure 8.12 The unsaturated intermediate is thought to be favoured in scCO$_2$ due to a carbonyl–CO$_2$ donor–acceptor interaction

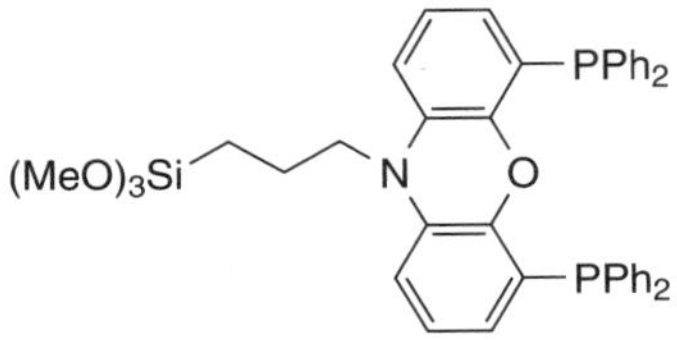

Figure 8.13 N-(3-trimethoxysilyl-n-propyl)-4,5-bis(diphenylphosphino)phenoxazine

good (as expected with this ligand which performs well in homogeneous conditions) although the rate of reaction was only moderate. Importantly, the catalyst was found to be very robust, with its performance being constant over at least 30 h with no rhodium leaching being found. The use of scCO$_2$ also facilitated product purification – about 90 % of unreacted alkene could be separated from the product by simply controlling a two-step depressurization of CO$_2$.

8.4   CONCLUSIONS

The use of alternative solvents in hydrogenation and hydroformylation reactions has developed at an incredible rate over the last few years. Many elegant systems have been designed which offer cleaner alternatives to those carried out in conventional organic solvents. Apart from the attractiveness of the separation process, catalyst lifetimes can be extended which represents another major advantage. In some cases, conventional organic solvents are completely removed from the system.

REFERENCES

1. Whyman R. *Applied Organometallic Chemistry and Catalysis*, Oxford University Press, Oxford, 2001.
2. Jourdant A., González-Zamora E. and Zhu J. *J. Org. Chem.* 2002, **67**, 3163.
3. McMurry J. *Organic Chemistry*, 4th Edition, Brooks/Cole, Pacific Grove, 1996, p. 542.
4. Dror Y. and Manassen J. *J. Mol. Catal.* 1977, **2**, 219.
5. (a) Okano T., Iwahara M., Suzuki T., Konishi H. and Kiji J. *Chem. Lett.* 1986, 1467; (b) Okano T., Iwahara M., Konishi H. and Kiji J. *J. Organomet. Chem.* 1988, **346**, 267.
6. Plasseraud L. and Süss-Fink G. *J. Organomet. Chem.* 1997, **539**, 163.
7. Widegren J. A. and Finke R. G. *J. Mol. Catal. A* 2003, **191**, 187.
8. (a) Januszkiewicz K. R. and Alper H. *Organometallics* 1983, **2**, 1055; (b) Januszkiewicz K. R. and Alper H. *Can. J. Chem.* 1984, **62**, 1031.
9. Suarez P. A. Z., Dullius J. E. L., Einloft S., de Souza R. F. and Dupont J. *Polyhedron* 1996, **15**, 1217.
10. Suarez P. A. Z., Dullius J. E. L., Einloft S., de Souza R. F. and Dupont J. *Inorg. Chim. Acta* 1997, **255**, 207.
11. Chauvin Y. and Olivier-Bourbigou H. *CHEMTECH* 1995, **25**, 26.

12. Berger A., de Souza R. F., Delgado M. R. and Dupont J. *Tetrahedron: Asymmetry* 2001, **12**, 1825.
13. Dyson P. J., Ellis D. J., Henderson W. and Laurenczy G. *Adv. Synth. Catal.* 2003, **345**, 216.
14. Dyson P. J., Ellis D. J. and Welton T. *Can. J. Chem.* 2001, **79**, 705.
15. Richter B., Deelman B.-J. and van Koten G. *J. Mol. Catal. A* 1999, **145**, 317.
16. Richter B., Spek A. L., van Koten G. and Deelman B.-J. *J. Am. Chem. Soc.* 2000, **122**, 3945.
17. de Wolf E., Spek A. L., Kuipers B. W. M., Philipse A. P., Meeldijk J. D., Bomans P. H. H., Frederik P. M., Deelman B.-J. and van Koten G. *Tetrahedron* 2002, **58**, 3922.
18. (a) Hitzler M. G. and Poliakoff M. *Chem. Commun.* 1997, 1667; (b) Hitzler M. G., Smail F. R., Ross S. K. and Poliakoff M. *Org. Proc. Res. Dev.* 1998, **2**, 137.
19. Burk M. J., Feng S., Gross M. F. and Tumas W. J. *J. Am. Chem. Soc.* 1995, **117**, 8277.
20. Leitner W. *Acc. Chem. Res.* 2002, **35**, 746.
21. Kuntz E. Fr. Pat. 2314910, 1975; DE 2627354.
22. Cornils B. and Kuntz E. G. *J. Organomet. Chem.* 1995, **502**, 177.
23. Wesserscheid P. and Waffenschmidt H. *J. Mol. Catal. A Chem.* 2000, **164**, 61.
24. Dickinson R. E. and Cicerone R. J. *Nature* 1986, **319**, 109.
25. Chauvin Y., Mussmann L. and Oliver H. *Angew. Chem., Int. Ed. Engl.* 1995, **34**, 2698.
26. Brasse C. C., Englert U., Salzer A., Waffenschmidt H. and Wesserscheid P. *Organometallics* 2000, **19**, 3818.
27. (a) Horváth I. T. and Rábai J. *Science* 1994, **266**, 72; (b) Horváth I. T., Kiss G., Cook R. A., Bond J. E., Stevens P. A. and Rábai J. *J. Am. Chem. Soc.* 1998, **120**, 3133.
28. (a) Foster D. F., Gudmunsen D., Adams D. J., Stuart A. M., Hope E. G. and Cole-Hamilton D. J. *Chem. Commun.* 2002, 722; (b) Foster D. F., Gudmunsen D., Adams D. J., Stuart A. M., Hope E. G., Cole-Hamilton D. J., Schwarz G. P. and Pogorzelec P. *Tetrahedron*, 2002, **58**, 3901.
29. (a) Frohling C. D. and Kohlpaintner C. W. In *Applied Homogeneous Catalysis with Organometallic Compounds*, Cornils B. and Herrmann W. A. (eds), VCH, Weinheim, 1996, Vol. 1, pp. 27–104; (b) MacDougall J. K., Simpson M. C., Green M. J. and Cole-Hamilton D. J. *J. Chem. Soc. Dalton Trans.* 1996, 1161.
30. Mathivet T., Monflier E., Castanet Y., Mortreux A. and Couturier J.-L. *Tetrahedron* 2002, **58**, 3877.
31. (a) van Leeuwen P. W. M. N. and Roobeck C.-F. *J. Organomet. Chem.* 1983, **258**, 343; (b) van Rooy A., Orij E.-N., Kramer P. C. J. and van Leeuwen P. W. N. M. *Organometallics* 1995, **14**, 34.
32. Billig E., Abatjoglou A. G., Bryant D. R., Murray R. E. and Mahler J. M. *Int. Pat.* WO 85/030702, 1985.
33. Bonafoux D., Hua Z., Wang B. and Ojima I. *J. Fluorine Chem.* 2001, **112**, 101.
34. Sakai N., Mano S., Nozaki K. and Takay H. *J. Am. Chem. Soc.* 1993, **115**, 7033.
35. Francio G., Wittmann K. and Leitner W. *J. Organomet. Chem.* 2001, **621**, 130.
36. Sellin M. F. and Cole-Hamilton D. J. *Chem. Commun.* 2000, 1681.
37. Hu Y., Chen W., Banet Osuna A. M., Stuart A. M., Hope E. G. and Xiao J. *Chem. Commun.* 2001, 725.
38. Meehan N. J., Sandee A. J., Reek J. N. H., Kamer P. C. J., van Leeuwen P. W. N. M. and Poliakoff M. *Chem. Commun.* 2000, 1497.

9   From Alkanes to CO$_2$: Oxidation in Alternative Reaction Media

Oxidation is the first step for producing molecules with a very wide range of functional groups because oxygenated compounds are precursors to many other products. For example, alcohols may be converted to ethers, esters, alkenes, and, via nucleophilic substitution, to halogenated or amine products. Ketones and aldehydes may be used in condensation reactions to form new C–C double bonds, epoxides may be ring opened to form diols and polymers, and, finally, carboxylic acids are routinely converted to esters, amides, acid chlorides and acid anhydrides. Oxidation reactions are some of the largest scale industrial processes in synthetic chemistry, and the production of alcohols, ketones, aldehydes, epoxides and carboxylic acids is performed on a mammoth scale. For example, world production of ethylene oxide is estimated at 58 million tonnes, 2 million tonnes of adipic acid are made, mainly as a precursor in the synthesis of nylons, and 8 million tonnes of terephthalic acid are produced each year, mainly for the production of poly(ethylene terephthalate) [1].

Scheme 9.1 shows a generalized sequence of reactions for the oxidation of an alkane, via alcohol, ketone and carboxylic acid, to the completely oxidized products, water and carbon dioxide. The latter are often referred to as *combustion products* as they are the same as those formed by burning hydrocarbons. These are not normally desirable chemical products unless it is necessary to destroy a toxic, hazardous or otherwise unwanted waste material. Oxidation itself is not difficult to achieve, and is a highly exothermic or even explosive process. Selective oxidation, however, is a much greater challenge, as it is important to stop the sequence at the desired product without proceeding further down the oxidation pathway.

Many laboratory and even some industrial scale oxidations were historically conducted using stoichiometric, toxic, metal-based oxidants such as KMnO$_4$, K$_2$Cr$_2$O$_7$ and OsO$_4$ [2]. However, the use of small-molecule sources of oxygen is preferable from both economic and environmental viewpoints. These oxidants include O$_2$, H$_2$O$_2$ and NaOCl, with an additional metal catalyst if required.

As a general rule, oxidation reactions produce products of greater polarity than the substrates, as shown in Table 9.1 [3]. These changes in polarity make oxidation reactions very suitable candidates for biphasic reactions. For example, if an alkene is oxidized to a more polar product (e.g. an epoxide or diol) in a

Chemistry in Alternative Reaction Media   D. Adams, P. Dyson and S. Tavener
© 2004 John Wiley & Sons, Ltd   ISBNs: 0-471-49848-3 (Cloth); 0-471-49849-1 (Paper)

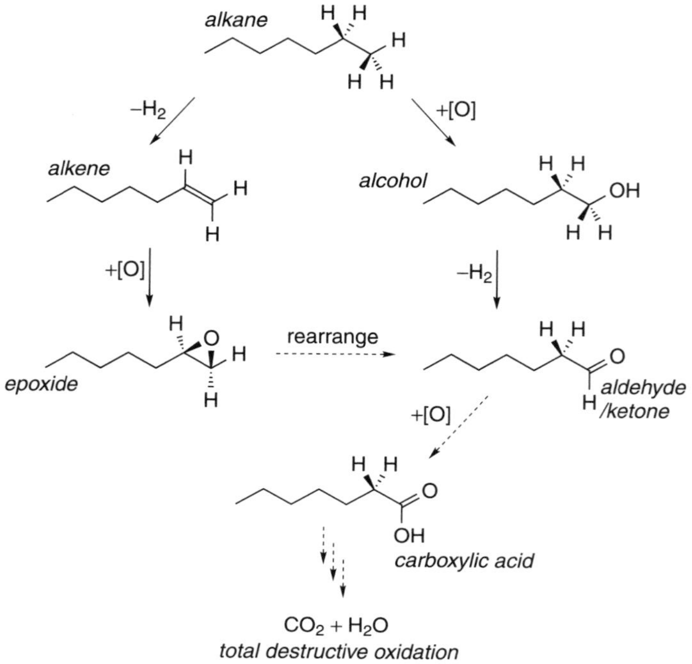

Scheme 9.1

Table 9.1 E_T^N and Kamlet–Taft $\pi*$ polarity parameters for n-hexane and oxidized derivatives

Compound	E_T^N	$\pi*$
n-Hexane	0.09	−0.04
n-Hexanol	0.50	0.40
2-Hexanone	0.27	0.65
n-Hexanoic acid	0.76	0.52

fluorous biphase, the product should be less soluble in the fluorous phase than the substrate, and this should lead to improved separation. In addition, oxygen has a remarkably high solubility in perfluorinated solvents (see Table 9.2), probably because neither the gas nor the solvent has strong intermolecular attractions. For these reasons, the fluorous biphase is an ideal medium in which to conduct oxidation reactions using elemental oxygen.

Ionic liquids based on imidazolium cations and either $[BF_4]^-$ or $[PF_6]^-$ anions have been used to immobilize transition metal based oxidation catalysts developed

Table 9.2 Solubility of oxygen in perfluorinated solvents[a]

Solvent	Solubility of O_2 (ml/100 ml solvent)
n-Perfluorooctane, C_8F_{18}	52.1
Perfluorotributylamine, $(C_4F_9)_3N$	38.4
Perfluorooctyl bromide, $C_8F_{17}Br$	52.7
FC-75[b]	52.2
Perfluorodecalin	40.3

[a] See Chapter 3 for more information about these solvents.
[b] Mainly perfluorobutyltetrahydrofuran.

for use in organic solvents. In general, these ionic liquids are stable to the reaction conditions, which include the use of oxidants such as O_2, H_2O_2 and NaOCl. However, $[PF_6]^-$ is known to break down under extreme conditions, forming HF, and care must be taken to ensure that the solvent is not degraded. Many oxidations have also been successfully performed in supercritical and aqueous systems. The rest of this chapter describes the oxidation reactions outlined in Scheme 9.1 conducted in alternative reaction media.

9.1 OXIDATION OF ALKANES

Oxidation of unfunctionalized alkanes is notoriously difficult to perform selectively, because breaking of a C–H bond is required. Although oxidation is thermodynamically favourable, there are limited kinetic pathways for reaction to occur. For most alkanes, the hydrogens are not labile, and, as the carbon atom cannot expand its valence electron shell beyond eight electrons, there is no mechanism for electrophilic or nucleophilic substitution short of using extreme (superacid or superbase) conditions. Alkane oxidations are therefore normally radical processes, and thus difficult to control in terms of selectivity. Nonetheless, some oxidations of alkanes have been performed under supercritical conditions, although it is probable that these actually proceed via radical mechanisms.

The oxidation of propane has been reported using a silica-supported Co_3O_4 catalyst (containing 2.4 % metal by weight) in $scCO_2$ (Scheme 9.2) [4]. Cobalt salts are effective catalysts for radical oxidations because the Co(II) and Co(III) oxidation states are similar in energy, which leads to a higher rate of radical initiation [5]. The reaction was performed as a mixture of substrate, air and CO_2 in the proportions 1 : 2.5 : 112, respectively. This gave rise to a range of oxygenated products (in total 59 % of the converted alkane) as well as some propene (21 %). The remaining 20 % probably represents substrate which has been destructively oxidized to H_2O and CO_2. Reducing the CO_2 density was found to improve the selectivity for the oxygenated products.

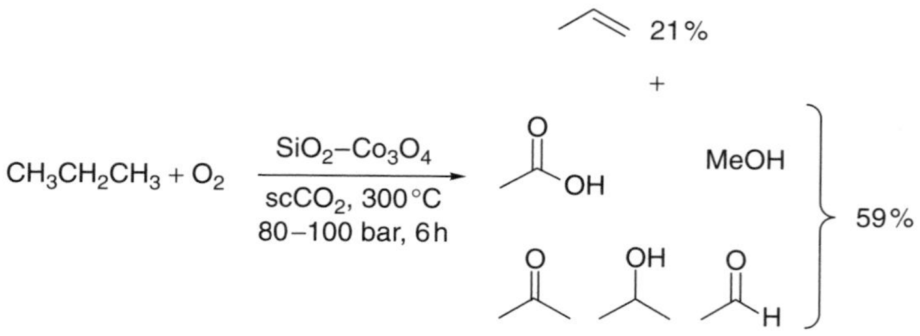

Scheme 9.2

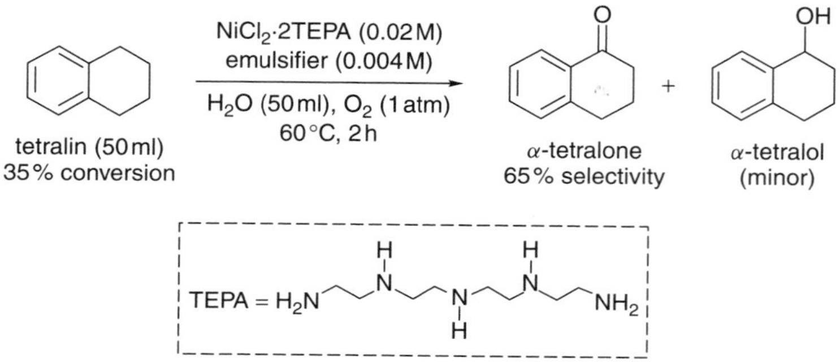

Scheme 9.3

Partial oxidation of n-hexadecane has been achieved under the more extreme conditions of scH_2O, but this gave a complex mixture of small (C1 to C4) hydrocarbons, as well as CO_2, CO and H_2 [6].

The conversion of tetralin to α-tetralone may be achieved under aqueous biphasic conditions in the presence of O_2, using $NiCl_2$ as a catalyst, as shown in Scheme 9.3 [7]. It is necessary to use tetraethylene pentamine (TEPA) as a surface-active ligand, as well as an emulsifier, dodecyl sodium sulfate. Some alcohol and naphthol by-products were also observed, and a radical chain mechanism has been proposed for this reaction.

9.2   OXIDATION OF ALKENES

The asymmetric epoxidation of unfunctionalized alkenes has been carried out in a fluorous biphase using a range of chiral perfluoroalkylated salen-type manganese complexes, including the one shown in Scheme 9.4 [8]. These catalysts were formed via reaction of preformed salen ligands with $Mn(OAc)_2$. Although many alkenes were examined, only indene gave the corresponding epoxide with high

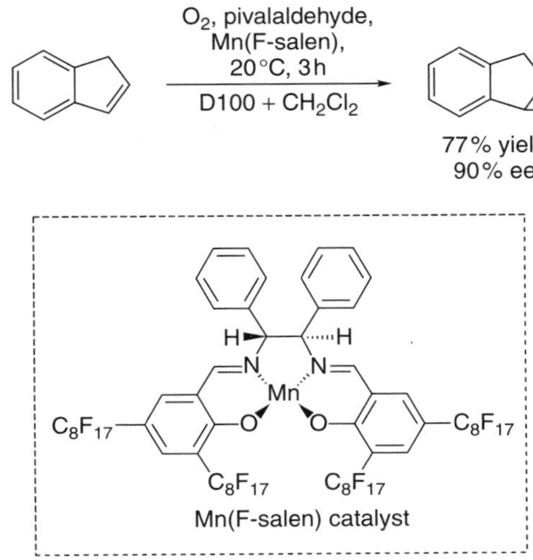

O₂, pivalaldehyde,
Mn(F-salen),
20°C, 3h
D100 + CH₂Cl₂

77% yield
90% ee

Mn(F-salen) catalyst

Scheme 9.4

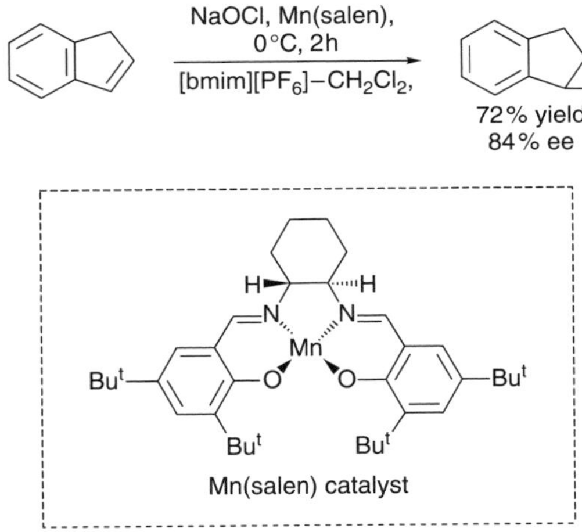

NaOCl, Mn(salen),
0°C, 2h
[bmim][PF₆]–CH₂Cl₂,

72% yield
84% ee

Mn(salen) catalyst

Scheme 9.5

enantioselectivities, and a sacrificial aldehyde was also required. Epoxidations in fluorous biphasic systems can also be catalysed by fluorinated tetraarylpor-phyrin–cobalt complexes under similar conditions [9].

Stereoselective alkene epoxidations have been performed using the ionic liquid [bmim][PF₆] in a biphase with dichloromethane, using a Mn-salen catalyst [10], as shown in Scheme 9.5. This gave yields in excess of 70 % and enantiomeric

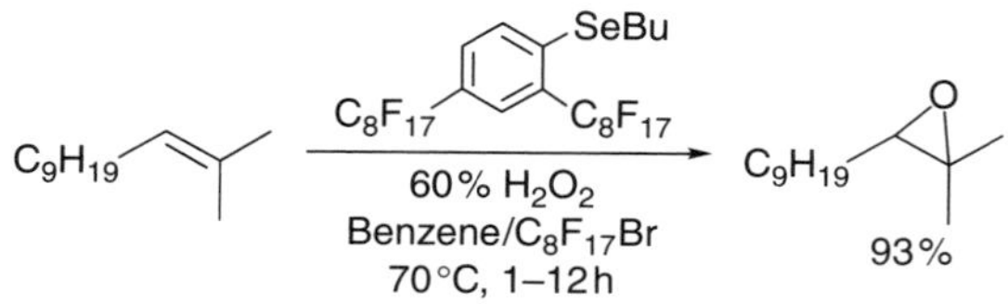

Scheme 9.6

excesses of up to 97 %, and the active phase could be recycled five times without loss of activity. The same salen ligand may be used with chromium to catalyse the ring opening of epoxides in [bmim][PF$_6$] and other ionic liquids [11]. Chiral ammonium salt phase transfer catalysts give impressive results when used for stereoselective epoxidations with NaOCl as the oxidant [12].

Selenium-catalysed epoxidations have been carried out using a perfluoroalkylated phenylselenide together with hydrogen peroxide, as shown in Scheme 9.6 [13]. The catalyst could be recycled 10 times with no loss of activity, with the toxic selenide immobilized in the perfluorinated solvent.

Hydrogen peroxide is a useful oxidizing agent, and, as it produces only water and O$_2$ as by-products, it has certain environmental advantages. The anthraquinone process, in which an alkyl anthraquinone is first hydrogenated and then oxidized, is used to supply almost all of the global demand for H$_2$O$_2$ [14]. The anthraquinone process may be successfully performed in scCO$_2$ if the anthraquinone catalyst is made compatible with the fluid phase by functionalization with perfluorinated chains [15]. Moreover, the H$_2$O$_2$ produced in this way may be utilized in the same reactor (i.e. a one-pot process), for the epoxidation of alkenes (Scheme 9.7). Hydrogen peroxide has also been used for epoxidation reactions in scCO$_2$ without the need for any added metal catalyst [16]. In these reactions, peroxycarbonic acid (HOCOOOH) is thought to be formed via reaction of CO$_2$ with H$_2$O$_2$, and then acts as an activated oxidizing agent. Hydrogen peroxide may also be used for the epoxidation of electrophilic alkenes such as 2-cyclohexen-1-one, in the ionic liquids [bmim][BF$_4$] and [bmim][PF$_6$]. Quantitative conversion occurs without the need for a transition metal catalyst [17].

Alkenes may be converted directly to ketones in a fluorous biphase via the *Wacker oxidation* [18]. The solvent system of benzene and C$_8$F$_{17}$Br is homogeneous at the reaction temperature of 56 °C, but separates into two phases on cooling to allow recycling of the fluorous phase. A palladium catalyst with a perfluorinated ligand is used, and *t*-butylhydroperoxide acts as the oxidant. A wide range of substituted styrenes and aliphatic alkenes react under these conditions, as shown in Scheme 9.8. The fluorous phase, including the catalyst, could be recycled with very little loss in activity. The yield fell from 78 % on the first run to 72 % on the eighth cycle.

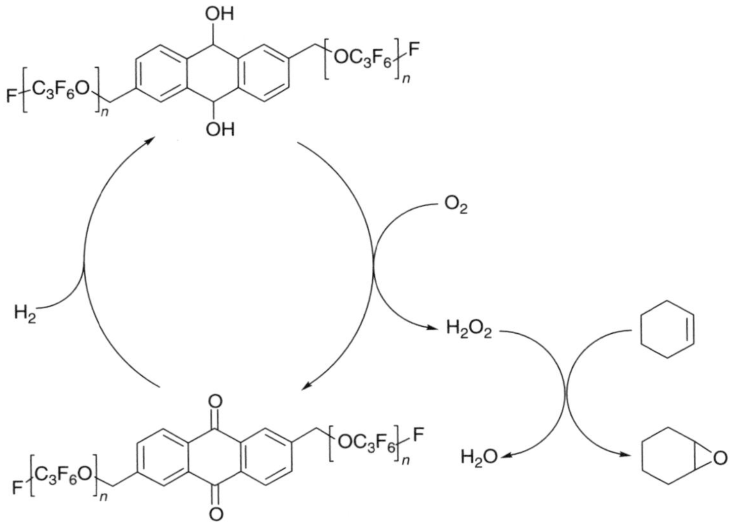

Scheme 9.7

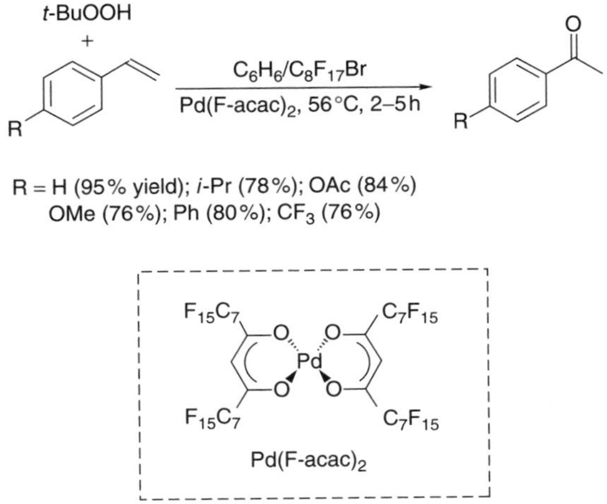

R = H (95% yield); *i*-Pr (78%); OAc (84%)
OMe (76%); Ph (80%); CF_3 (76%)

Pd(F-acac)$_2$

Scheme 9.8

Wacker oxidation of styrene has also been performed in [bmim][BF$_4$] and [bmim][PF$_6$], at 60 °C with H_2O_2 and PdCl$_2$ as a catalyst [19]. This system gave yields of acetophenone as high as 92% after 3 h. Hydrogen peroxide may also be used under phase transfer conditions for alkene bond cleavage, to produce adipic acid (an intermediate in the synthesis of nylon-6) from cyclohexene (Scheme 9.9).

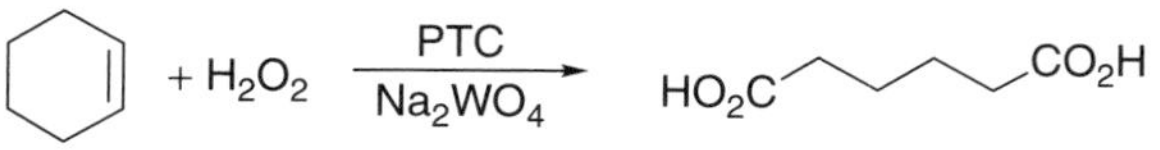

Scheme 9.9

9.3 OXIDATION OF ALCOHOLS

The oxidation of primary and secondary alcohols to the corresponding aldehydes and ketones without over-oxidation to carboxylic acids is an important process in organic synthesis due to their versatility as functional groups. An organic nitrogen oxide, 2,2,6,6-tetramethylpiperidinyl-1-oxy (TEMPO) has been found to be an extremely effective catalyst for selective oxidations and, in conventional solvents, it is used in combination with stoichiometric amounts of appropriate oxidants [20]. In the ionic liquid [bmim][PF$_6$], a TEMPO–CuCl catalyst has been used to oxidize primary and secondary alcohols to aldehydes and ketones with O$_2$ as the oxidant, as shown in Scheme 9.10 [21]. Benzylic and allylic alcohols are oxidized with excellent conversion, although aliphatic alcohols are not very reactive under these conditions. The selectivity is good, with no detectable quantities of the acid produced. The reaction media can be reused a number of times with only a slight decrease in activity, with the yield decreasing from 72 % on the first run to 60 % on the eighth run for the oxidation of benzyl alcohol to benzaldehyde.

A similar reaction has been conducted under fluorous biphasic conditions, using a perfluoroalkylated bipyridine as ligand to ensure that the copper species resides in the fluorous phase [22]. The oxidation of a range of primary alcohols to the corresponding aldehydes was found to be possible, an example of which is shown in Scheme 9.11. The catalyst could be successfully recycled by phase separation, with analytically pure products being isolated even after

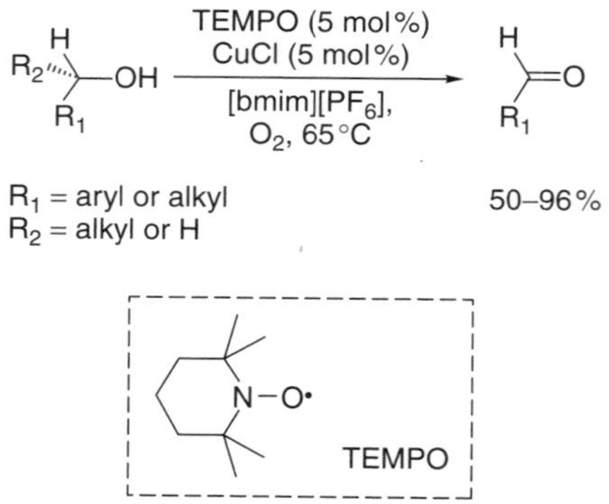

R_1 = aryl or alkyl   50–96 %
R_2 = alkyl or H

Scheme 9.10

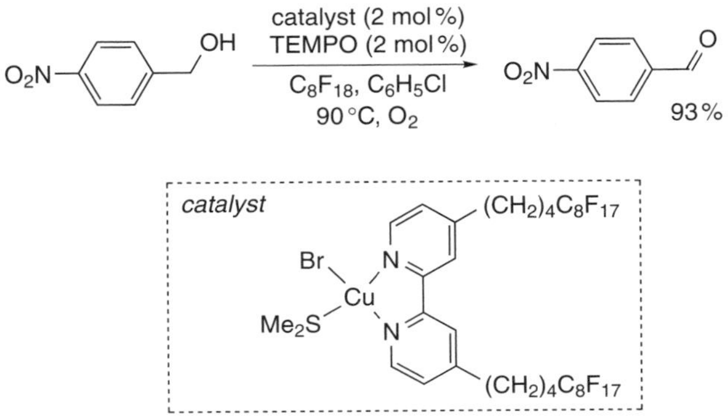

Scheme 9.11

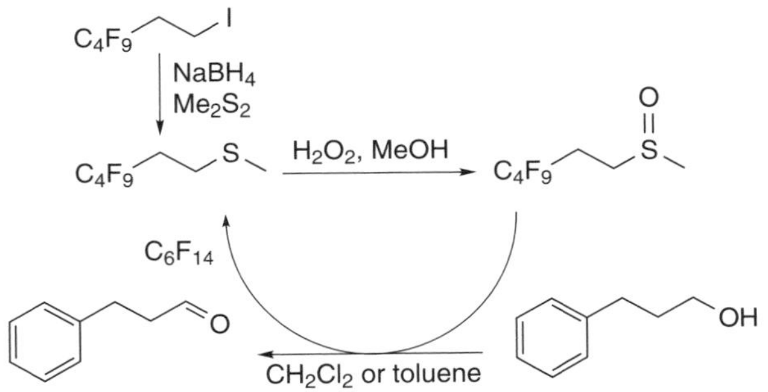

Scheme 9.12

eight cycles. Alcohols have also been oxidized to aldehydes and ketones in [bmim][$(CF_3SO_2)_2N$] by using [Pr_4N][RuO_4] and CuCl catalysts and air as the source of oxygen [23].

A fluorous analogue of DMSO has been used to perform *Swern* reactions [24]. This widely used method of oxidizing an alcohol to an aldehyde falls down seriously from the environmental point of view due to its production of a stoichiometric amount of dimethyl sulfide. Here, a fluorous sulfoxide is prepared and used in the oxidation of several alcohols in dichloromethane, as shown in Scheme 9.12. After reaction, the sulfide is extracted into perfluorohexane and the system recycled. Unfortunately, extraction from dichloromethane was found to be difficult, but replacing the dichloromethane with toluene leads to a more efficient recovery.

The enzyme *cholesterol oxidase* is an effective oxidation catalyst when used in a micellular system involving $scCO_2$ and water [25]. The use of a perfluoropoly-ether-based surfactant causes the formation of reverse-micelles in which the

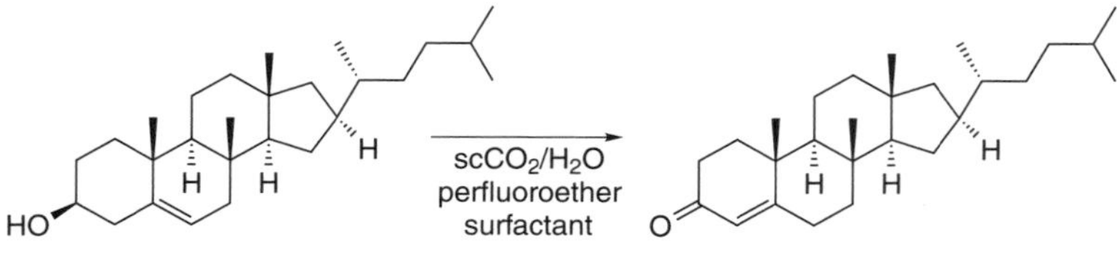

Scheme 9.13

enzyme is active. Reaction rates observed for the oxidation of cholesterol to cholestenone, as shown in Scheme 9.13, were twice as fast as those measured for a comparable system using a liquid isooctane and water solvents. Enzymes are also active in ionic liquids, and lipase *CaLB* will oxidize cyclohexene to cyclohexene oxide in [bmim][BF$_4$] at room temperature in the presence of H$_2$O$_2$ [26]. A yield of 83 % was obtained after 24 h.

9.4   OXIDATION OF ALDEHYDES AND KETONES

Aldehydes may be converted to carboxylic acids using Ni(acac)$_2$ immobilized in [bmim][PF$_6$], with oxygen as the oxidant, as shown in Scheme 9.14 [27]. A similar reaction has also been performed using perfluorinated solvents, and it was found that there was little difference between the two systems [28]. However, the Ni(acac)$_2$ catalyst could not be used directly in the fluorous solvent and therefore the 1,3-diketonate was modified with long perfluorinated chains prior to use to ensure solubility.

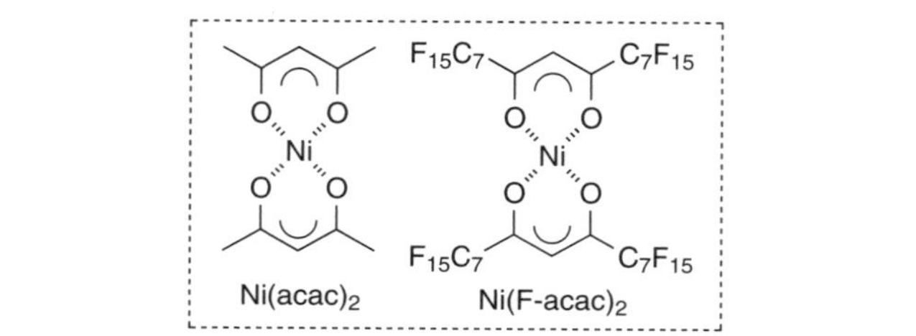

Scheme 9.14

In both the $Ni(acac)_2-[bmim][PF_6]$ and fluorous biphasic systems, catalyst leaching is very low and several further batch oxidation reactions may be carried out with similar results to those obtained in the first run. In the fluorous biphasic system, the yield of 4-chlorobenzoic acid dropped from 87 % to 70 % by the sixth reaction cycle; using $[bmim][PF_6]$ as a solvent, the yield was essentially the same after four uses, and no catalyst was found to leach into the organic phase.

Oxidation reactions are not limited to those that occur at a carbon centre. The perfluorinated $Ni(F-acac)_2-benzene-C_8F_{17}Br$ system described above was also active for the oxidation of sulfides to sulfoxides and sulfones [28]. A sacrificial aldehyde is required as co-reductant, but the reaction may be tuned by changing the quantity of this aldehyde. If 1.6 equivalents of aldehyde are used, the sulfoxide is obtained, whereas higher quantities (5 equivalents) lead to sulfones. Fluorous-soluble transition metal porphyrin complexes also catalyse the oxidation of sulfides in the presence of oxygen and 2,2-dimethylpropanal [29].

9.5 DESTRUCTIVE OXIDATION

Supercritical water is an excellent medium for the destruction of unwanted organic compounds [30]. Oxidation in scH_2O takes place at much lower temperatures than incineration in air, the most common method currently used. Unlike water at lower temperatures and pressures, organic compounds are often soluble in scH_2O. Conversely, inorganic compounds become almost insoluble. The concentrations of H^+ and OH^- are much higher than in liquid water, and in the presence of O_2 highly reactive hydroxyl radicals are formed [30, 31]. Combined with the high solubilities of gases and high diffusivities obtainable (common for all SCFs, see Chapter 6), these factors make it is possible to achieve very high rates of oxidation for many organic compounds. Oxidation in scH_2O may initially result in the formation of smaller compounds, which are then further oxidized to form carbon dioxide and water. For example, Scheme 9.15 shows some of the steps in the sequential oxidation of phenol in scH_2O [32]. Elements such as chlorine and sulfur are normally converted to stable inorganic anions (i.e. Cl^- and SO_4^-), which makes this an attractive way to dispose of hazardous waste materials. Examples of destructive oxidation in scH_2O include chlorinated aromatics [33], waste chemicals from naval applications consisting mainly of kerosene and chlorinated solvents [34], aqueous effluent from poly(ethylene terephthalate) production [35], and even pet food, as a model for household waste [36].

Chlorinated organic compounds may be destroyed, but these produce acidic HCl which may corrode the reactor. Addition of Na_2CO_3 to these reactions not only reduces the corrosiveness of the mixture, but also enhances the rate of oxidation [37]. This is possibly because the Na_2CO_3 precipitates out from the SCF as a fine suspension, providing a large surface on which reaction may occur.

Supercritical water systems are capable of destroying any organic compound, even those with low biodegradability, which could be a useful way of disposing

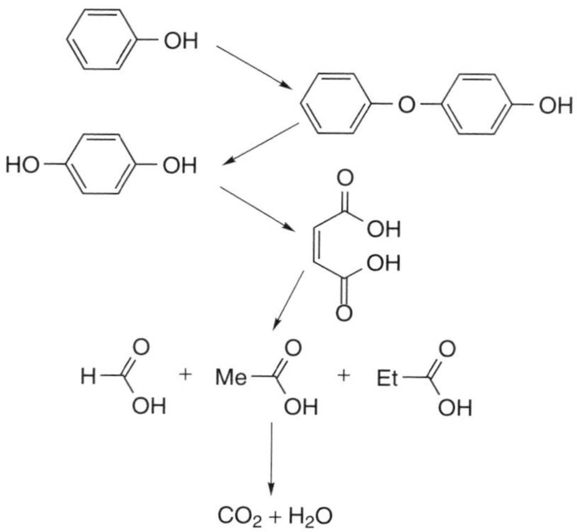

Scheme 9.15

of waste plastics, or even recycling them back to monomers or other useful small-chain hydrocarbons [38]. This has been applied to a range of polymers including polystyrene, poly(vinyl chloride) and polypropylene [39]. Under supercritical conditions, the polymers swell and may partially dissolve, leading to pseudo-homogeneous conditions. By controlling the amount of oxygen present, the polymer may either be decomposed by chain scission, giving methane, ethene, propene, *iso*-butene, propane, butane, benzene, toluene, ethylbenzene and styrene, or else total oxidation may occur at higher oxygen concentrations. Rubber tyres have also been converted to a fuel oil in 44 % yield by reaction in scH_2O at 400 °C, in a process that also removes sulfur from the product [40].

9.6 CONCLUSIONS

A wide range of oxidation reactions may be performed in alternative media. Small-molecule sources of oxygen, including O_2 and H_2O_2 may be utilized, and, where a metal catalyst is required, this may often be recycled using a biphasic approach. Both of these factors are desirable from an efficiency point of view, as the quantity of waste products is minimized. Fluorous solvents are particularly well suited to oxidation reactions using molecular O_2 because of its high solubility in these solvents, although ionic liquids and SCFs have also produced impressive results. Biphasic systems involving water as one of the phases (i.e. aqueous biphasic, phase transfer, and $scCO_2$–aqueous) are effective when aqueous H_2O_2 is required as the oxidant, and scH_2O is useful for the destructive oxidation of unwanted or hazardous waste chemicals.

REFERENCES

1. Weissermel K. and Arpe H.-J., *Industrial Organic Chemistry*, 2nd Edition, VCH, Weinheim, 1993.
2. (a) Clark J. H. *Green Chem.* 1999, **1**, 1; (b) Sheldon R. A. *J. Chem. Tech. Biotechnol.* 1997, **68**, 381.
3. Marcus Y. *Chem. Soc. Rev.* 1993, 409.
4. Kerler B. and Martin A. *Catal. Today* 2000, **61**, 9.
5. Suresh A. K., Sharma M. M. and Sridhar T. *Ind. Eng. Chem. Res.* 2000, **39**, 3958.
6. Watanabe M., Mochiduki M., Sawamoto S., Adschiri T. and Arai K. *J. Supercrit. Fluids* 2001, **20**, 257.
7. Chung Y. M., Ahn W. S. and Lim P. K. *J. Catal.* 1998, **173**, 210.
8. Pozzi G., Cavazzini M., Cinato F., Montanari F. and Quici S. *Eur. J. Org. Chem.* 1999, 1947.
9. Pozzi G., Montanari F. and Quici S. *Chem. Commun.* 1997, 69.
10. Song C. E. and Roh E. J. *Chem. Commun.* 2000, 837.
11. Song C. E., Oh C. R., Roh E. J. and Choo D. J. *Chem. Commun.* 2000, 1743.
12. Lygo B. and To D. C. M. *Tetrahedron Lett.* 2001, **42**, 1343.
13. Betzemeier B., Lhermitte F. and Knochel P. *Synlett* 1999, 489.
14. Sanderson W. R. In *Handbook of Green Chemistry and Technology*, Clark J. H. and Macquarrie D. J. (eds), Blackwell, Oxford, 2002, p. 256.
15. Hâncu D., Green J. and Beckman E. J. *Acc. Chem. Res.* 2002, **35**, 757.
16. Nolen S. A., Lu J., Brown J. S., Pollet P., Eason B. C., Griffith K. N., Glaser R., Bush D., Lamb D. R., Liotta C. L., Eckert C. A., Thiele G. F. and Bartels K. A. *Ind. Eng. Chem. Res.* 2002, **41**, 316.
17. Bortolini O., Conta V., Chiappe C., Fantin G., Fogagnolo M. and Maietti S. *Green Chem.* 2002, **4**, 94.
18. Betzemeier B., Lhermitte F. and Knochel P. *Tetrahedron Lett.* 1998, **39**, 6667.
19. Nanboodiri V. V., Varma R. S., Sahle-Demessie E. and Pillai U. R. *Green Chem.* 2002, **4**, 170.
20. (a) Semmelhack M. F., Schnid C. R., Cortes D. A. and Chou C. S. *J. Am. Chem. Soc.* 1984, **106**, 3374; (b) Dijksman A., Arendes I. W. C. E. and Sheldon R. A. *Chem. Commun.* 1999, 1951.
21. Ansar I. A. and Gree R. *Org. Lett.* 2002, **4**, 1507.
22. Betzemeier B., Cavazzini M., Quici S. and Knochel P. *Tetrahedron Lett.* 2000, **41**, 4343.
23. Farmer V. and Welton T. *Green Chem.* 2002, **4**, 97.
24. (a) Crich D. and Neelamkavil S. *J. Am. Chem. Soc.* 2001, **123**, 7449; (b) Crich D. and Neelamkavil S. *Tetrahedron* 2002, **58**, 3865.
25. Kane M. A., Baker G. A., Pandey S. and Bright F. V. *Langmuir* 2000, **16**, 4901.
26. Sheldon R. A., Lau R. M., Sorgedrager M. J., van Rantwijk F. and Seddon K. R. *Green Chem.* 2002, **4**, 147.
27. Howarth J. *Tetrahedron Lett.* 2000, **41**, 6627.
28. Klement I., Leutjens H. and Knochel P. *Angew. Chem., Int. Ed. Engl.* 1997, **36**, 1454.
29. Colonna S., Gaggero N., Montanari F., Pozzi G. and Quici S. *Eur. J. Org. Chem.* 2001, 181.
30. Savage P. E. *Chem. Rev.* 1999, **99**, 603.
31. Feng J., Aki S. N. V. K., Chateauneuf J. E. and Brennecke J. F. *J. Am. Chem. Soc.* 2002, **124**, 6302.
32. Krajnc M. and Levec J. *AIChE J.* 1996, **42**, 1977.
33. Clifford A. A. In *Chemistry of Waste Minimization*, Clark J. H. (ed.), Chapman & Hall, London, 1995, p. 504.

34. Crooker P. J., Ahluwalla K. S., Fan Z. and Prince J. *Ind. Eng. Chem. Res.* 2000, **39**, 4865.
35. Cocero M. J., Alonso E., Torio R., Vallelado D., Sanz T. and Fdz-Polanco F. *Ind. Eng. Chem. Res.* 2000, **39**, 4652.
36. Mizuno T., Goto M., Kodama A. and Hirose T. *Ind. Eng. Chem. Res.* 2000, **39**, 2807.
37. Muthukumaran P. and Gupta R. B. *Ind. Eng. Chem. Res.* 2000, **39**, 4555.
38. Aguado J. and Serrano D. *Feedstock Recycling of Plastic Wastes*, Royal Society of Chemistry, Cambridge, 1999.
39. Lee S., Gencer M. A., Fullerton K. L. and Azzam F. O. US Pat. 5516952, 1996.
40. Park S. and Gloyna E. F. *Fuel* 1997, **76**, 999.

10 Carbon–Carbon Bond Formation, Metathesis and Polymerization

In Chapters 8 and 9, various catalysed reactions involving gas substrates (hydrogenation, hydroformylation and oxidation) that have been evaluated in alternative solvent media were described. In this chapter, we move onto reactions in which gaseous substrates are not usually used, these being C–C and C=C bond forming reactions, olefin (alkene) metathesis reactions and polymerizations. The formation of new single and double carbon–carbon bonds is extremely important in the synthesis of many valuable chemical products. Olefin metathesis represents a relatively new way in which to make highly valuable products which could only otherwise be made by much more complicated routes. The last type of reactions discussed in this chapter are polymerizations, and polymer products have a huge range of applications from inexpensive plastic bags to polymers that are able to conduct electricity. This chapter concentrates on areas where alternative solvent technologies can make positive contributions in terms of reactivity, product purity or processing, but also highlights potential problems and limitations.

10.1 CARBON–CARBON COUPLING REACTIONS

One of the most important objectives in organic synthesis, especially important in the synthesis of fine chemical products such as pharmaceuticals, is the facile synthesis of new C–C bonds. While C–C bond formation is relatively easy for certain substrates, it remains a major problem when retention of a double bond is required or an aromatic carbon atom is involved. Many different types for C–C bond forming reactions are known, notable examples being Heck, Suzuki, Stille and Sonogashira Negishi coupling reactions, which share a common feature that the catalyst used is based on palladium. While these reactions were originally developed in conventional organic solvents, many have been evaluated in alternative solvents and some examples are discussed below.

Chemistry in Alternative Reaction Media   D. Adams, P. Dyson and S. Tavener
© 2004 John Wiley & Sons, Ltd   ISBNs: 0-471-49848-3 (Cloth); 0-471-49849-1 (Paper)

10.1.1 Heck Coupling Reactions

A schematic of the Heck reaction is shown in Scheme 10.1, although it is worth noting that many variations on the Heck reaction, such as intramolecular and tandem Heck reactions, are known.

Although the Heck reaction is synthetically very useful, it requires quite high molar quantities of palladium catalyst to be effective. As such, one of the main goals is to find a solvent that helps to increase the lifetime of the catalyst and consequently reduce the amount of catalyst required. In this respect, ionic liquids show considerable promise. Another key goal in this area is to be able to replace iodo- and bromoarenes, usually used as substrates in these reactions, with chloroarenes, which are more environmentally acceptable. Again, ionic liquids show some promise in this respect. Scheme 10.2 shows the Heck reaction between styrene and chlorobenzene that has been investigated in a number of ionic liquids.

In fact, a large number of different catalysts, co-catalysts, bases and ionic liquids have been investigated for this reaction [1]. The most effective system takes place in [NBu$_4$]Br and uses the palladium(0) compound Pd$_2$(dba)$_3$ (dba = dibenzylideneacetone) with P(^{t}Bu)$_3$, and Na(OAc) base. The three dba ligands weakly coordinate to the two palladium(0) centres and are displaced under the reaction conditions by the bulky phosphine thereby generating the active catalyst. This combination gave a yield of 92 % in 18 h at 150 °C. The analogous reaction was examined in DMF, previously shown to be an excellent solvent for Heck reactions, and a yield of only 72 % was obtained. The use of Pd$_2$(dba)$_3$ without a phosphine ligand results in very low yields, and phosphine ligands less bulky than P(^{t}Bu)$_3$ also reduce the yield.

When Heck reactions and other C–C coupling reactions are carried out in imidazolium-based ionic liquids, the base can react with the acidic proton on the

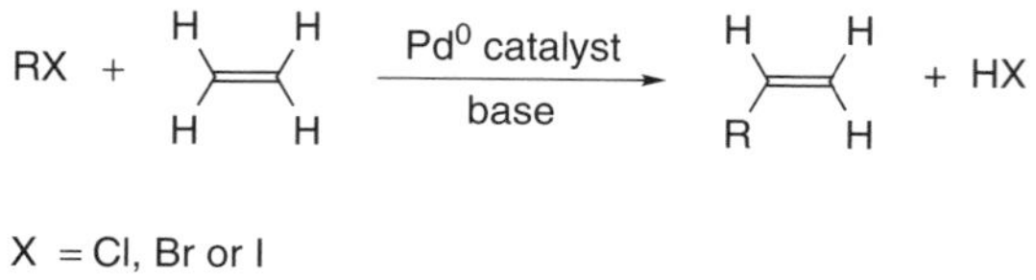

X = Cl, Br or I

Scheme 10.1

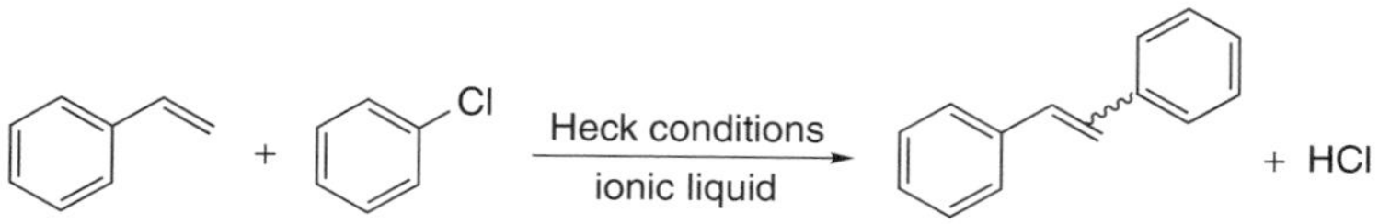

Scheme 10.2

2-position of the imidazolium cation to form carbenes [2]. It has even been possible to isolate the carbene complex cis-Pt(η-C$_2$H$_4$)(1-ethyl-3methylimidazol-2-ylidene)Cl$_2$ in moderate yield from the reaction of PtCl$_2$ and PtCl$_4$ with ethylene in [emim]-AlCl$_3$ ionic liquid (see Figure 10.1) [3].

The problem with carbene formation is that they can displace the phosphine ligands attached to the catalyst and deactivate the catalyst. In general, the active catalyst is a palladium(0) compound and this low oxidation state is best stabilized by very bulky phosphines such as P(^{t}Bu)$_3$ mentioned above.

Palladium-catalysed Heck reactions have been carried out in a mixture of acetonitrile and D-100 as shown in Scheme 10.3. Using a fluorinated phosphine allowed simple recycling of the catalyst by decantation [4].

A fluorous version of the chiral bis-phosphine BINAP has been developed for asymmetric Heck reactions [5]. Several fluorous-derivatized binaphthols and BINAP derivatives have been reported, (Scheme 10.4) [6]. The silane spacer group present in one of the ligands was used to maximize the percentage fluorine on the molecule. Even so, the partition coefficient[1] between FC-72 (see Chapter 3) and benzene was only 2.85, and not surprisingly, the reuse of the catalyst was poor.

Fluorous-derivatized phosphines have also been used for Heck reactions in scCO$_2$ together with palladium acetate [7]. The fluorous groups improve the solubility of the catalyst in the SCF compared to nonfluorous ligands. An example of a Heck reaction that uses a fluorous-derivatized phosphine to improve the solubility of a Pd(OAc)$_2$ catalyst is shown in Scheme 10.5.

The reaction between iodobenzene and methyl acrylate in scCO$_2$ using the Pd(OAc)$_2$/fluorous-derivatized phosphine catalyst gave a superior yield of methyl cinnamate compared to the same reaction in a conventional organic solvent. The

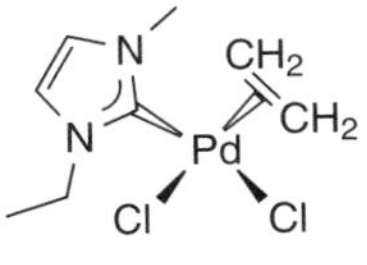

Figure 10.1   Structure of the carbene complex cis-Pt(η-C$_2$H$_4$)(1-ethyl-3-methylimidazol-2-ylidene)Cl$_2$

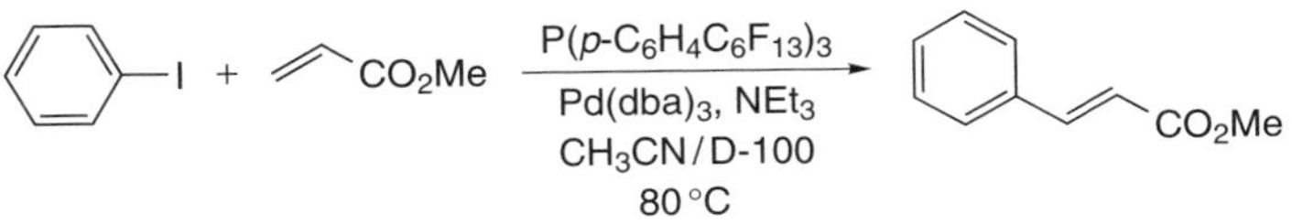

Scheme 10.3

---

[1] See Chapter 2 for definition of partition coefficients.

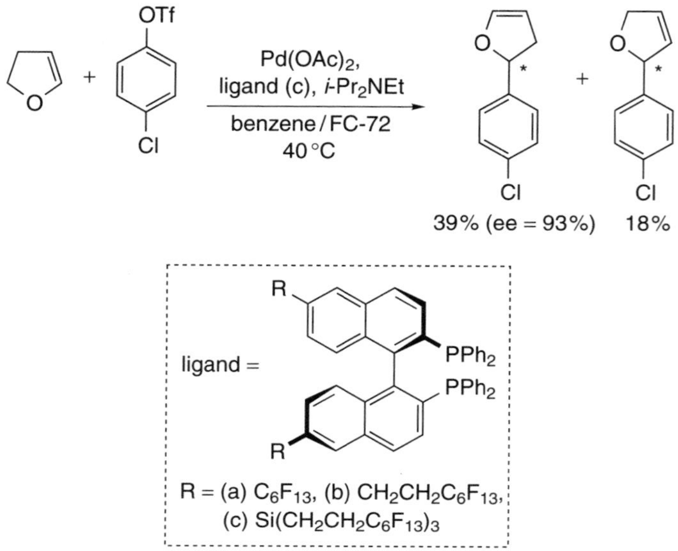

Scheme 10.4

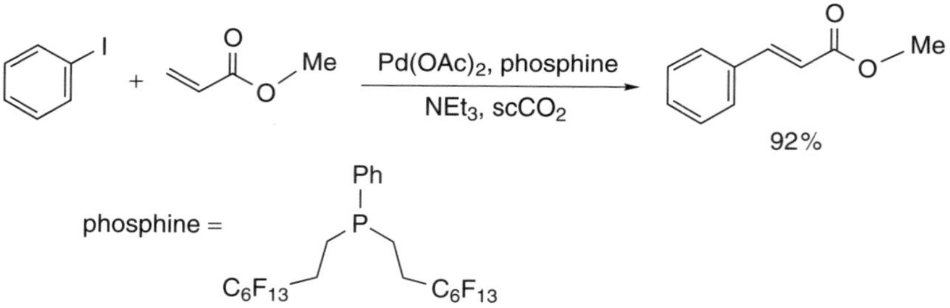

Scheme 10.5

work-up procedure for the reaction was significantly easier than that required under conventional reaction conditions.

Although the use of fluorous-derivatized ligands is very effective in solubilizing the catalyst in scCO$_2$, there would be advantages in using commercially available ligands in their place, as the derivatization step would be eliminated. Use of commercially available phosphines can be achieved by careful choice of the initial palladium source. A series of palladium compounds has been investigated, including several fluorinated species [8]. Palladium trifluoroacetate and palladium hexafluoracetylacetanoate were found to be very active catalysts when combined with commercially available phosphines, even allowing lower catalyst

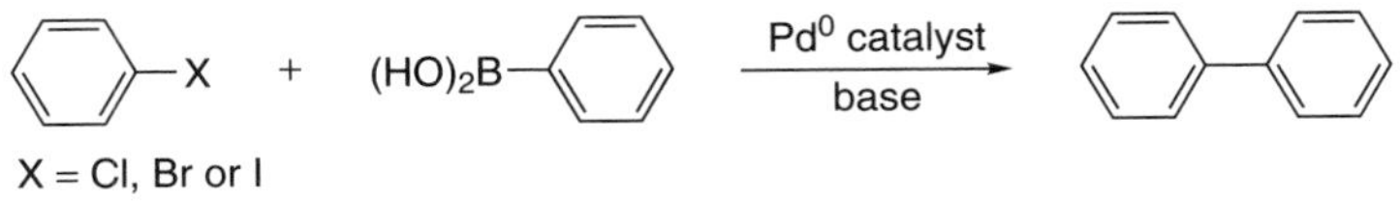

X = Cl, Br or I

Scheme 10.6

loadings to be used. With a moderate increase in catalyst loading, the reaction even proceeded in the absence of ligands.

10.1.2 Suzuki Coupling Reactions

As mentioned above, there are many different types of palladium catalysed reactions that lead to the formation of new carbon–carbon bonds. Stille reactions, for example, are very effective, but involve coupling between aryl halides and organotin reagents. There are some concerns surrounding the use of tin compounds and the detrimental effect they can have on the environment. Fortunately, good recycling and reuse of the tin can be achieved. The Suzuki reaction is perceived to be a more benign reaction as it involves the coupling between an aryl halide and an aryl boronic acid (or sometimes non-aryl boronic acids or boronates). A schematic of the Suzuki reaction is shown in Scheme 10.6.

A number of Suzuki reactions (see Scheme 10.9) have been conducted in ionic liquids using $Pd(PPh_3)_4$ as the catalyst at $30\,^{\circ}C$ [10]. Although the catalyst is neutral, the ionic liquid–catalyst solution can be used repeatedly without a decrease in activity. In fact, the catalyst shows a significant increase in activity compared to when it is used in conventional organic solvents. Another attractive feature of the system is that $NaHCO_3$ and $Na[XB(OH)_2]$ (X = halide) by-products can be removed from the ionic liquid–catalyst phase by washing with water.

Palladium nanoparticles coated with a fluorous-derivatized surfactant (or stabilizer) have been used to catalyse Heck and Suzuki reactions in a $C_8F_{17}Br$–benzene biphase [11]. The reaction between phenylboronic acid and cinnamyl bromide was investigated and the reaction is illustrated in Scheme 10.10.

The palladium nanoparticle is prepared from the reaction of the stabilizer, 4, 4′-bis(perfluorooctyl)dibenzylideneacetone with palladium(II) chloride. The average size of the nanoparticle varied according the ratio of $PdCl_2$ to the stabilizer, but was typically around 4 or 5 nm. The initial yield observed in the Suzuki coupling reaction was 90 %, but decreased to 78 % after five consecutive runs. Fluorous boronates (alternative precursors in Suzuki reactions), have also been developed for use in fluorous biphasic processes [12]. A generic structure of a fluorous boronate is shown in Figure 10.2.

The fluorous boronate is highly soluble in fluorous solvents, but when the nonfluorous R-group combines with another nonfluorous R-group the resulting product becomes preferentially soluble in the organic phase into which it is automatically extracted.

Friedel–Crafts Reactions i-Ionic Liquids

The presence of Lewis acidic species in chloroaluminate ionic liquids has also been used to bring about various acid catalysed transformations that do not require additional catalysts. For example, acidic ionic liquids are ideally suited to Friedel–Crafts acylation reactions. In a traditional Friedel–Crafts acylation an acylium ion is generated by reaction between acyl chloride and $AlCl_3$ or $FeCl_3$ as shown in Scheme 10.7.

Acidic chloroaluminate ionic liquids are able to generate acylium ions and are therefore ideally suited to Friedel–Crafts reactions. Acylation of mono-substituted aromatic compounds in acidic chloroaluminate ionic liquids leads almost exclusively to substitution at the 4-position on the ring [9] (Scheme 10.8).

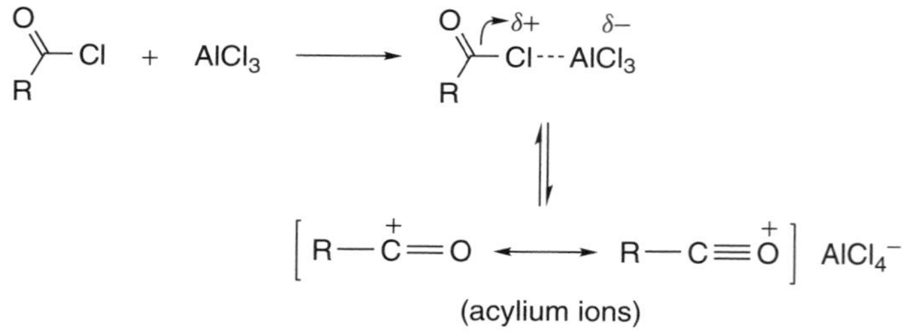

Scheme 10.7

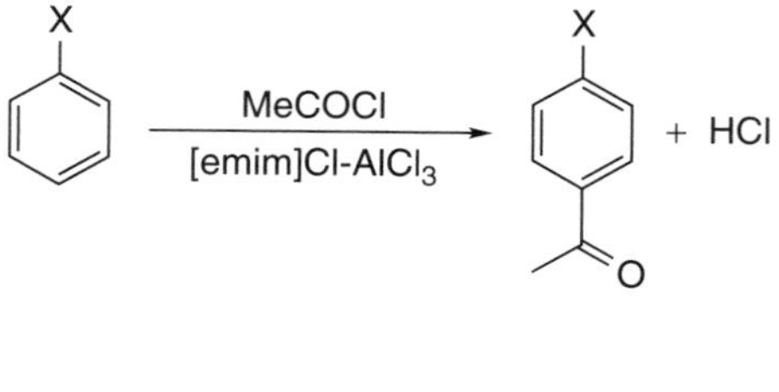

Scheme 10.8

10.1.3   Reactions Involving the Formation of C=C Double Bonds

The fluorous biphasic technique has also been applied to the Wittig reaction, one of the most effective methodologies for the production of C=C double bonds [13] (Scheme 10.11). One of the major drawbacks of this reaction is the separation of the alkene from the phosphine oxide by-product. This is commonly achieved via recrystallization or column chromatography, but recently it has been shown

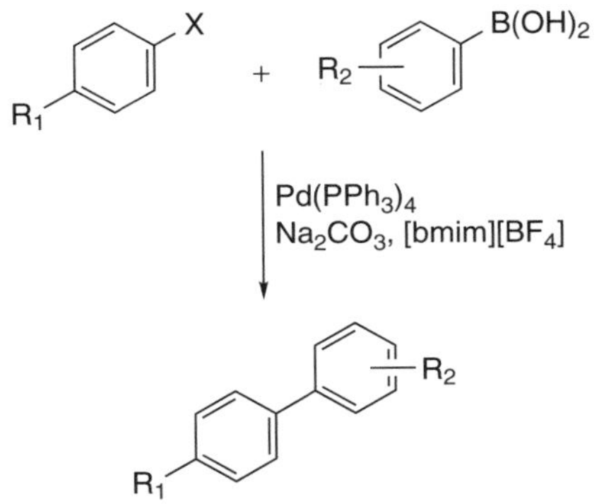

Scheme 10.9

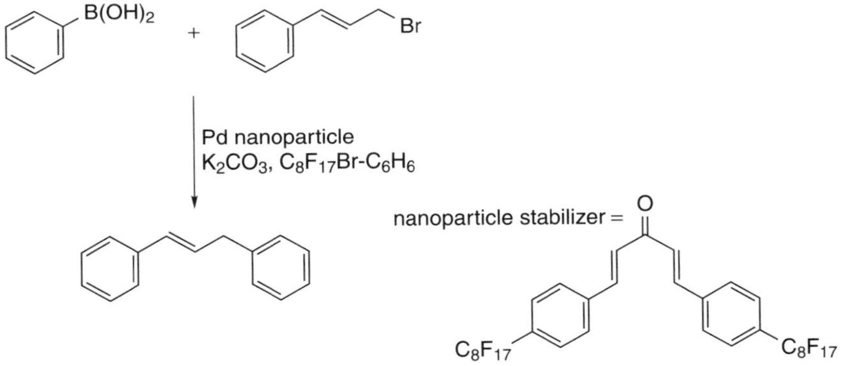

Scheme 10.10

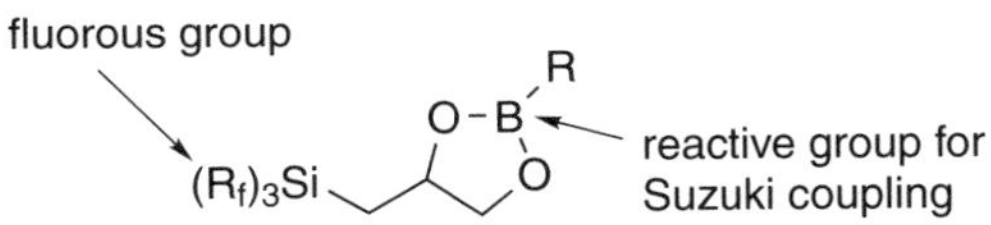

Figure 10.2   General structure of fluorous boronates used in fluorous biphase Suzuki reactions

that fluorinated phosphine oxides could be recycled quite effectively using a fluorous biphasic system. A number of Wittig reactions were carried out in a D-100–toluene biphase. At the end of the reaction, the product was isolated pure of any phosphine oxide residues by phase separation.

Since the Wittig reaction involves ionic intermediates it is ideally suited to the ionic liquid environment. Wittig reactions using stabilized ylides have been

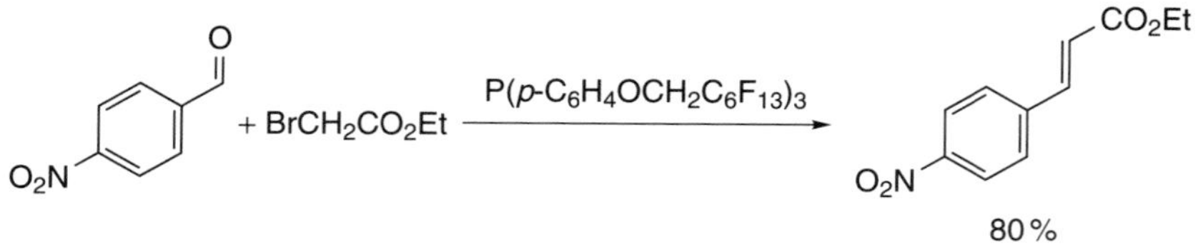

Scheme 10.11

conducted in [bmim][BF$_4$] giving conversions of typically 80–90 % [14]. After reaction the product can be separated from the ionic liquid and Ph$_3$PO by-product by extraction with diethyl ether followed by filtration through silica. The Ph$_3$PO is then extracted into toluene leaving clean ionic liquid for further reaction.

10.2 METATHESIS REACTIONS

Olefin (alkene) metathesis is an important reaction in which all the carbon–carbon double bonds in an alkene are cut and then rearranged in a statistical fashion as shown in Scheme 10.12.

The mechanism involves a [2 + 2] cycloaddition reaction between an alkene and a transition metal carbene (Scheme 10.13). In the absence of a transition metal carbene catalyst, the reaction between two alkenes is symmetry forbidden and only takes place photochemically. However, the d-orbitals on the metal catalyst (typically Grubbs's catalyst as shown in Scheme 10.13), break the symmetry and the reaction is facile.

10.2.1 Ring Opening Metathesis Polymerization

Ring opening metathesis polymerization (ROMP) is derived from the olefin metathesis reaction. It uses strained cyclic alkenes that react to form stereoregular polymers and copolymers. ROMP is often referred to as a *living polymerization* method as the polymer assembles in a controlled and stoichiometric way allowing a high level of control over the molecular weight. The polydispersities obtained in ROMP are typically in the range 1.03 to 1.10, which is so narrow the polymers

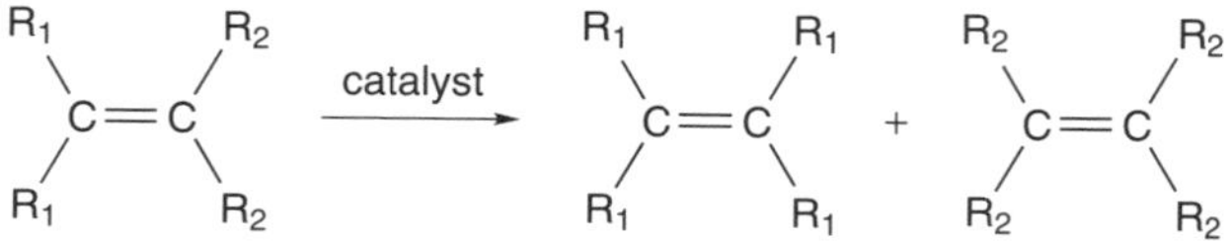

Scheme 10.12

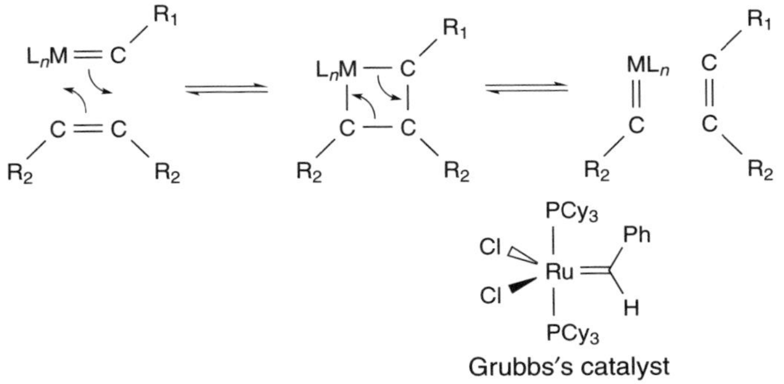

PCy$_3$

Cl ⟍ | Ph
⟍ Ru ═⟨
Cl ◥ | H
PCy$_3$

Grubbs's catalyst

Scheme 10.13

are said to be monodisperse. The ROMP mechanism is related to that of olefin metathesis, but with two notable differences:

1 The catalyst remains bonded to the new alkene formed from the cyclic alkene.
2 The driving force for the reaction is relief of the ring strain and the reaction is therefore irreversible and only works for strained alkenes.

The modified ROMP mechanism is summarized in Scheme 10.14.

ROMP reactions have been extensively carried out in water, and the first examples in liquid and scCO$_2$ [15] and ionic liquids [16] have been demonstrated. ROMP of norbornene and cyclooctene in scCO$_2$ exhibit a similar efficiency to that of chlorinated organic solvents. However, the carbon dioxide based system allows simple and highly convenient work-up of the polymer products. In [bdmim][PF$_6$] (bdmim = 1-butyl-2,3-dimethylimidazolium cation), norbornene has been polymerized using a cationic catalyst as shown in Scheme 10.15.

The best results, in terms of catalyst recycling, were obtained when toluene was used as a co-solvent with the ionic liquid. Without toluene, the yield of polymer produced decreases dramatically to only 10 % in the third cycle compared to 98 % in the first.

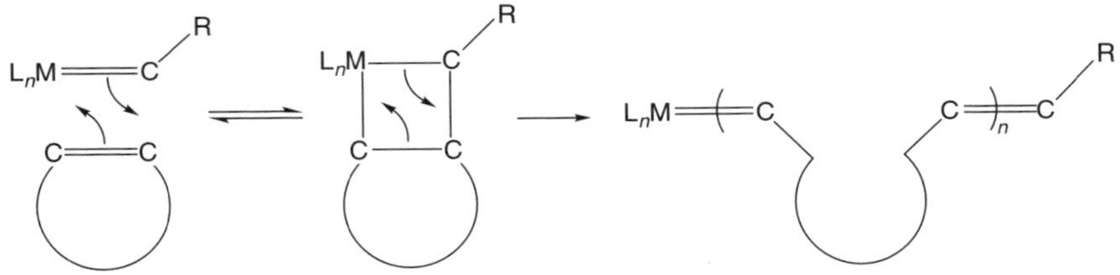

Scheme 10.14

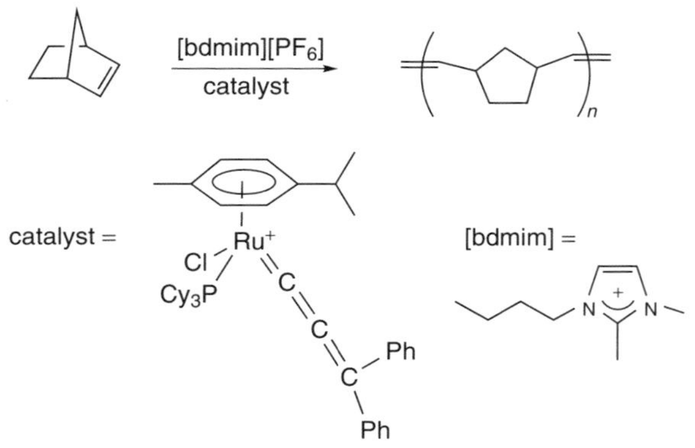

Scheme 10.15

10.2.2 Ring Closing Metathesis

Ring closing metathesis (RCM) is the reverse of ROMP. The reaction will only work if the ring being formed is largely unstrained. RCM has become a powerful method for the synthesis of heterocycles with ring sizes of five and above. RCM is often most effective under high dilution conditions which can necessitate the use of large volumes of solvent, and therefore finding cleaner alternatives to volatile organic solvents is important. RCM reactions have been carried out in water, ionic liquids, and $scCO_2$ (using $scCO_2$ under essentially the same conditions employed for ROMP). In RCM the driving force of the reaction is often the liberation of a volatile by-product, and therefore attempting the reaction under supercritical pressures would seem counter-intuitive. However, in the RCM reaction shown in Scheme 10.16, which liberates ethene, the reaction proceeds in almost quantitative yield under mild conditions.

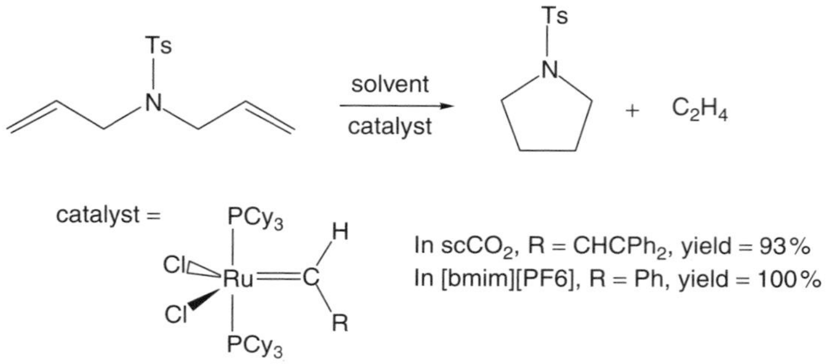

In $scCO_2$, R = $CHCPh_2$, yield = 93%
In [bmim][PF6], R = Ph, yield = 100%

Scheme 10.16

The same reaction has been achieved in 100 % yield in the ionic liquid [bmim][PF$_6$] after just 1 h. However, some ruthenium residues are detected in very low amounts in the product.

Thus far, considerably more research has been directed towards RCM in water. The majority of metathesis catalysts decompose rapidly in the presence of water or oxygen, however, Grubbs's ruthenium based catalysts are quite robust. Replacement of the tricyclohexylphosphine ligands with water soluble phosphines has allowed their deployment in aqueous–organic biphasic processes although conversions are often not as good as those obtained in other solvents [18].

10.3 POLYMERIZATION REACTIONS IN ALTERNATIVE REACTION MEDIA

Single and double carbon–carbon bond forming reactions, and metathesis (ROMP) reactions, can be used to produce polymers. However, many other

---

A Biphasic RCM/Heck Process

A fluorous biphasic system has been successfully used to carry out a bimetallic cascade ring closing metathesis and intramolecular Heck reaction [17]. Phosphines strongly retard RCM reactions. So, for a cascade reaction to be successful, the palladium catalyst with modifying phosphine ligands for the Heck reaction must be kept away from the other reagents until the ring closing is complete. Although this can be achieved using a polystyrene bound triphenylphosphine, a fluorous biphasic system can also be used. Here, the ring closing reaction occurs at room temperature in the organic solvent, with the palladium catalyst immobilized in the perfluorinated solvent using a perfluoroalkyl-derivatized triphenylphosphine. After the ring closing is complete, the system is heated and becomes monophasic, allowing the intramolecular Heck reaction to occur (Scheme 10.17).

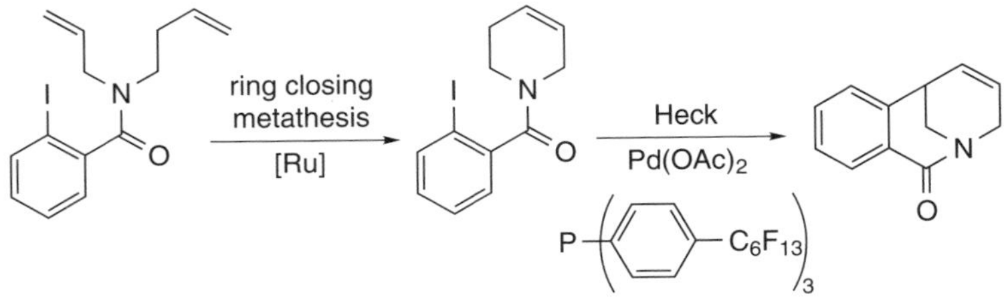

Scheme 10.17

types of reactions also lead to the formation of polymers, and since they are such important products, their preparation in alternative solvents using a wide range of different reactions will be described in this section.

10.3.1 Polymerization Reactions in Water

Water offers a number of important properties as a solvent for polymerization reactions. As well as its high polarity, which gives a markedly different miscibility with many monomers and polymers compared to organic solvents, it is nonflammable, nontoxic and cheap. Water also has a very high heat capacity that sustains heat exchanges in a number of very exothermic polymerizations. Largely because of these factors, polymerizations are now widely carried out in aqueous media, and, for example, more than 50 % of industrial radical polymerizations are carried out in water [19].

Free-radical initiated emulsion polymerization is carried out on a multimillion ton scale [20]. A typical polymerization mixture consists of water, a water-soluble initiator, a water-immiscible monomer and a suitable surfactant (or stabilizer). The stabilizer is present to stop the polymer particles coagulating and precipitating. This can be achieved electrostatically (with an ionic surfactant adsorbed onto a particle surface) or sterically (by covalently binding a water-soluble polymer to the particle surface). The initiator (e.g. $K_2S_2O_8$) forms a water-soluble radical (e.g. $SO_4^{\bullet-}$), which grows by addition of monomers. This water-soluble oligomeric radical continues to grow until it reaches a critical chain length where it becomes insoluble and collapses on itself. Once insoluble, it is stabilized by the surfactant and further polymerization occurs within the primary particle with further monomer diffusing into this particle. This emulsion polymerization results in a stable dispersion of polymer particles in water (otherwise known as a *latex*), and by altering the conditions, polymer spheres of different sizes may be produced (Figures 10.3 and 10.4).

In contrast to the free-radical polymerizations, there have been relatively few studies on transition metal catalysed polymerization reactions in water. This is largely due to the fact that the early transition metal catalysts used commercially for the polymerization of olefins tend to be very water-sensitive. However, with the development of late transition metal catalysts for olefin polymerizations, water is beginning to be exploited as a medium for this type of polymerization reaction. For example, cationic Pd(II)−bisphosphine complexes have been found to be active catalysts for olefin−CO copolymerization [21]. Solubility of the catalyst in water is achieved by using a sulfonated phosphine ligand (Figure 10.5) as described in Chapter 5.

Many other metal-catalysed polymerizations may be carried out in water including the copper-catalysed polymerization of methacrylates, the palladium- and nickel-catalysed polymerization of ethene and other alkenes and the rhodium-catalysed polymerization of butadiene [22].

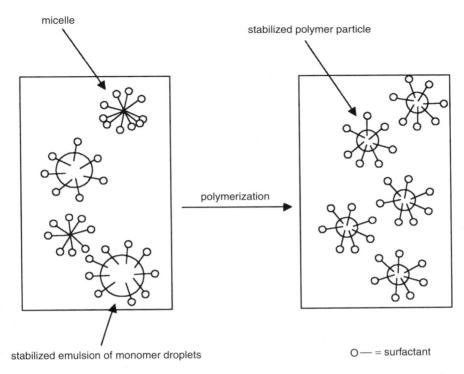

micelle

stabilized polymer particle

polymerization

stabilized emulsion of monomer droplets

O— = surfactant

Figure 10.3 A water-insoluble monomer is stabilized by a surfactant and polymerized to give a polymer latex

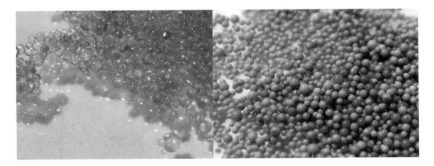

Figure 10.4 (Plate 8) Polystyrene spheres prepared by emulsion polymerization methods. Because they may be packed together to form columns or beds, these spheres find applications in separations, ion exchange, and as supports for catalysts. (Photographs by John Olive)

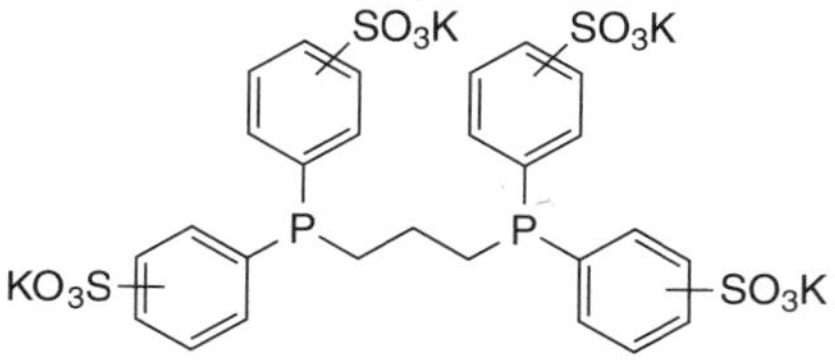

Figure 10.5   A diphosphine ligand used to render a polymerization catalyst water-soluble

10.3.1.1   Polymerizations under phase transfer conditions

Phase-transfer techniques are widely used for the preparation of polymers. For example, potassium fluoride is used to produce poly(etherketone)s under phase-transfer conditions (Scheme 10.18). Use of this reagent allows the chloroaromatics to be used as starting material as opposed to the more expensive fluoroaromatics that are usually employed [23]. This method is suitable for the synthesis of high molecular weight semicrystalline poly(ether ketone)s, although the presence of excess potassium fluoride in the reaction mixture can lead to degradation reactions. The use of a phase transfer catalyst can allow the use of water-soluble radical initiators, such as potassium peroxomonosulfate used to promote the free-radical polymerization of acrylonitrile [24].

10.3.2   Polymerization Reactions in Supercritical Carbon Dioxide

One of the first reports of polymerization under supercritical conditions was that of the high-pressure production of low-density polyethylene by ICI in the 1930s.

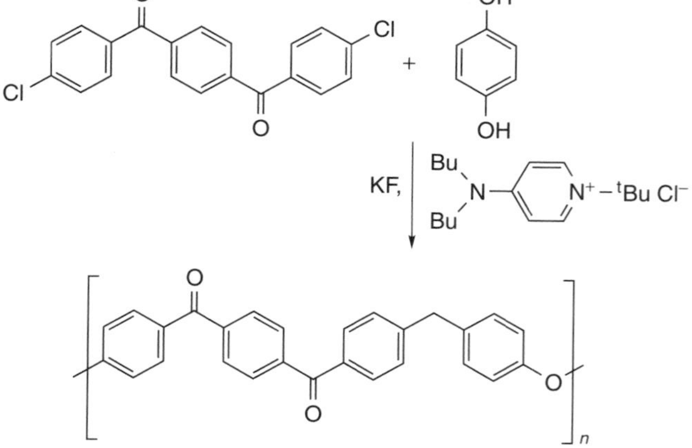

Scheme 10.18

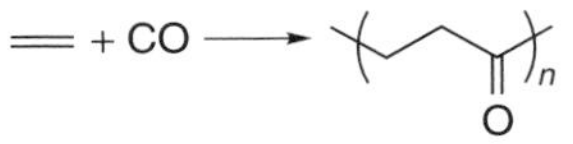

Scheme 10.19

This industrial process remains essentially unchanged from the 1950s [25]. Here, a free-radical initiator is added to the ethylene monomer at supercritical conditions (276 MPa and 200–300 °C). The polyethylene remains in the supercritical solution until the pressure is lowered to around 5 MPa, whereupon it precipitates. A range of other monomers can be copolymerized, including carbon monoxide to give polyketones, as shown in Scheme 10.19 [26].

Supercritical fluids have also been used purely as the solvent for polymerization reactions. Supercritical fluids have many advantages over other solvents for both the synthesis and processing of materials (see Chapter 6), and there are a number of factors that make scCO$_2$ a desirable solvent for carrying out polymerization reactions. As well as being cheap, nontoxic and nonflammable, separation of the solvent from the product is achieved simply by depressurization. This eliminates the energy-intensive drying steps that are normally required after the reaction. Carbon dioxide is also chemically relatively inert and hence can be used for a wide variety of reactions. For example, CO$_2$ is inert towards free radicals and this can be important in polymerization reactions since there is then no chain transfer to the solvent. This means that solvent incorporation into the polymer does not take place, giving a purer material.

One potential problem with using scCO$_2$ as a reaction medium for polymerization reactions is the low solubility of many polymers in it. Carbon dioxide is a good solvent for many nonpolar compounds with low molecular weights, but is a very poor solvent for most high molecular weight polymers. The only polymers that show good solubility in CO$_2$ under easily obtainable conditions are certain fluoropolymers and silicones. This lack of solubility does impose certain restrictions, but it has also been exploited in many processing applications.

10.3.2.1 Polymerization of fluorinated monomers

Amorphous fluoropolymers have many applications in the areas of advanced materials where they are used in applications requiring thermal and chemical resistance. Their manufacture is hindered by their low solubility in many solvents. Many fluoropolymerizations cannot be carried out in hydrocarbon solvents because the radical abstraction of hydrogen atoms leads to detrimental side reactions. Chlorofluorocarbons (CFCs) were thus commonly used, but their use is now strictly controlled due to their ozone depleting and greenhouse gas properties. Supercritical carbon dioxide is a very attractive alternative to CFCs and it has been shown that amorphous fluoropolymers can be synthesized by

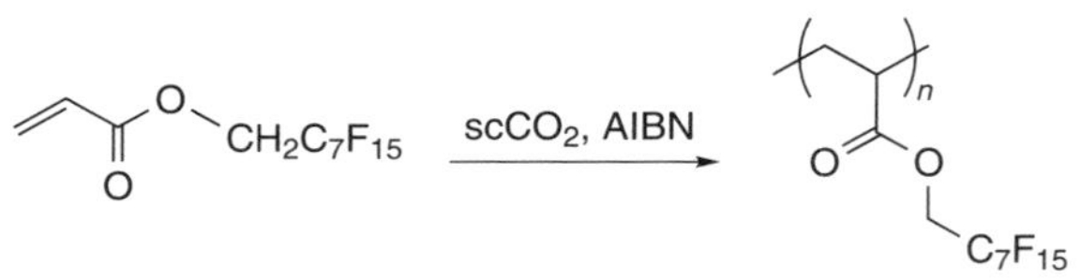

Scheme 10.20

homogeneous solution polymerization in $scCO_2$ [27]. For example, $1H, 1H$-perfluorooctylmethyl acrylate may be polymerized in $scCO_2$ using $2, 2'$-azobis(isobutyronitrile) (AIBN) as the initiator, as shown in Scheme 10.20.

Other fluorinated acrylates, methacrylates and styrenes may be polymerized or copolymerized in carbon dioxide [28]. 1,1-Difluoroethylene can be telomerized and the molecular weight and molecular weight distribution depend on the density of the $scCO_2$ [29]. DuPont have recently built a US\$275 million plant capable of manufacturing 1000 tonnes of Teflon^{TM} and other fluoropolymers per year. The plant uses CO_2 technology to produce a grade of polymer that DuPont claim has enhanced performance and processing capabilities, whilst generating less waste during manufacture [30]. Tetrafluoroethylene polymerization may also been carried out in CO_2. Here, the potentially explosive monomer is stored safely as a pseudo-azeotrope with CO_2 [31]. Copolymerizations with hexafluoropropene have been shown to give polymers with levels of co-monomer similar to those of commercial processes [32].

10.3.2.2 Heterogeneous polymerizations in supercritical carbon dioxide

Whilst $scCO_2$ is an extremely attractive alternative solvent for carrying out the homogeneous polymerization of fluorinated monomers, the low solubility of many high molecular weight polymers restricts the synthesis of such materials to heterogeneous techniques. This is not necessarily a problem, since a lack of solubility is a requirement for such techniques as suspension, emulsion and dispersion polymerizations, as described for water above. Many industrially important vinyl monomers have been polymerized in $scCO_2$ via free-radical precipitation techniques [33]. Free-radical precipitation polymerization of acrylic acids have also been successfully carried out in $scCO_2$. Very high molecular weight polymers ($M_n \approx 150 \times 10^3$ g mol^{-1}) can be synthesized despite the polymer precipitating from the solution [34]. Styrene [35], methyl methacrylate [36] and other monomers [37] have also been successfully polymerized and copolymerized.

Transition metal catalysed precipitation polymerization utilizes the advances in the development of suitable catalysts for homogeneous reactions in $scCO_2$. For example, the synthesis of poly(phenyl acetylene) may been achieved using a

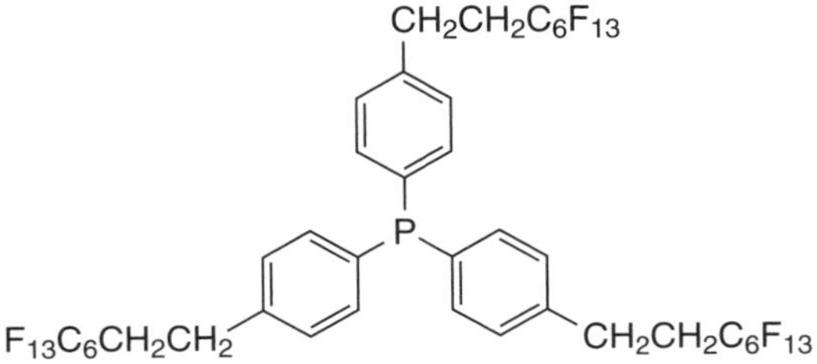

Figure 10.6 Substituted triphenylphosphine used to solubilize [Rh(nbd)(acac)] (nbd = norbornadiene, acac = acetylacetonate) precursor for the synthesis of poly(phenyl acetylene) in scCO$_2$

rhodium catalyst in either subcritical CO$_2$ or scCO$_2$. Although the catalyst precursor is insoluble in carbon dioxide, solubilization was achieved as for many organic reactions by the addition of a perfluoroalkyl-substituted triphenylphosphine ligand [38], shown in Figure 10.6.

Carbon dioxide can itself be used as a feedstock as well as a solvent for the synthesis of aliphatic polycarbonates by precipitation polymerization. Propylene oxide [39] and 1,2-cyclohexene oxide [40] can both be polymerized with CO$_2$ using a heterogeneous zinc catalyst (Scheme 10.21).

The polycarbonate precipitates from the scCO$_2$ as it forms. Although the weight averaged molecular masses achieved were fairly high ($M_w = (25–150) \times 10^3$ g mol^{-1}), the molecular weight distributions were broad (1.9–3.4) and the catalyst efficiencies were low. However, by using a CO$_2$-soluble catalyst the catalyst efficiencies were improved [41].

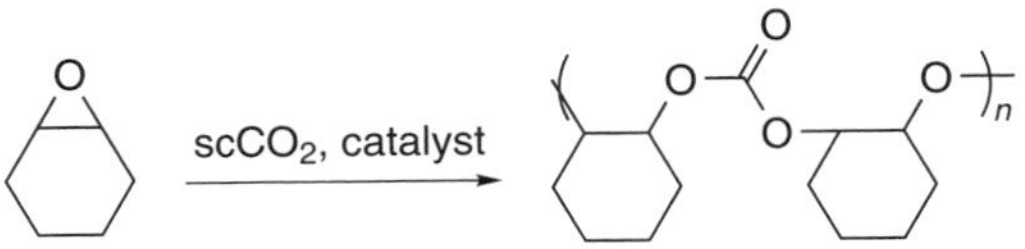

Scheme 10.21

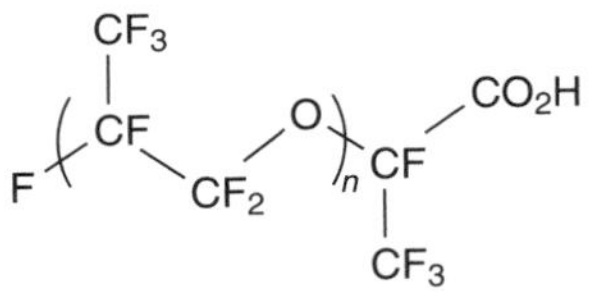

Figure 10.7 Krytox 157 FSL ($M_n = 2500$, $n \approx 14$)

Polymer Moulding Using Supercritical Fluids

Cross-linked macroporous monoliths made from polymers are attracting interest as continuous porous separation media. Typically, such monoliths are prepared by filling a mould with a cross-linking monomer, comonomer, initiator and a suitable porogenic solvent. The mixture is then polymerized, taking up the shape of the mould. Polymers with well-defined pore structures can be produced by careful control of the nature of the solvent as well as monomer and initiator concentrations. One disadvantage of this methodology is the large volume of organic solvents required in their production. However, such monoliths can be made using $scCO_2$ as the solvent [45]. For example, the copolymerization of two acrylic monomers has been carried out, and after reaction, the CO_2 is vented, leaving a continuous porous monolith which conformed with the shape of the reaction vessel (Figure 10.8). It is even possible to fine-tune the pore sizes by varying the density of CO_2 used.

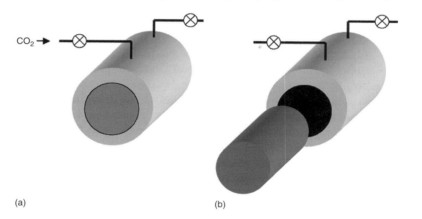

(a)   (b)

Figure 10.8 The monomers are polymerized inside a high-pressure view cell (a). After reaction, the $scCO_2$ is vented and the monolith removed from the cell (b)

10.3.2.3 Dispersion polymerization

In dispersion polymerization, homogeneous conditions exist at the start of the reaction, but the resulting polymer is insoluble and hence phase separation occurs, producing a heterogeneous system. Common solvents for polymerization of lipophilic monomers such as styrene include hydrocarbons or low molecular weight alcohols. A surfactant (or stabilizer) is required for successful polymerization. Without the stabilizer, the particle dispersions are not sufficiently stable and tend to coagulate. Examples include methacrylic polymers with oligostearic grafts or a polar organic polymer such as poly(vinylpyrrolidone). Under the correct conditions, dispersion polymerization gives rise to well-defined,

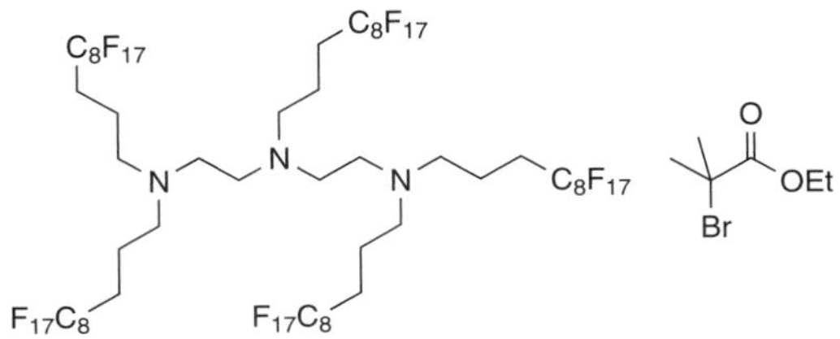

Figure 10.9

spherical particles, typically between $0.1–10\,\mu m$ in diameter. Although $scCO_2$ has the potential to replace the large volumes of organic solvents currently used, the major disadvantage with $scCO_2$ is that common surfactants and stabilizers used for these methodologies are insoluble or ineffective [42], and so expensive alternatives (typically perfluorinated surfactants) are required. For a stabilizer to be effective in $scCO_2$, it must contain both a CO_2-phobic and a CO_2-philic moiety. Poly(1,1-dihydroperfluorooctylacrylate), shown as the product in Scheme 10.20, has been used as such a stabilizer with considerable success. For example, the free-radical dispersion polymerization of methyl methacrylate in $scCO_2$ alone gives a polymer with a relatively low molecular weight ($77–149 \times 10^3\,g\,mol^{-1}$) with low monomer conversion (10–40%) [43]. However, when 2–4% w/v of poly(*1H,1H*-perfluorooctylmethyl acrylate) is also included in the reaction, the phase behaviour was found to be very different, with a stable, opaque-white colloidal dispersion formed. Both the monomer conversion (>90%) and the product molecular weights ($190–325 \times 10^3\,g\,mol^{-1}$) are much improved. Importantly, upon venting the CO_2, poly(methyl methacrylate) can be recovered as a dry, free-flowing powder which consists of uniform spherical particles with average diameters in the range $1.2–2.5\,\mu m$. One drawback of this type of approach is that the stabilizer often ends up in the polymer. This is because the stabilizer contains a hydrocarbon backbone from which protons may be abstracted during free-radical polymerizations, leading to incorporation into the growing monomer. Hence, a different type of stabilizer has also been used. A carboxylic acid terminated perfluoropolyether, Krytox 157 FSL (Figure 10.7) has been chosen because a H-bond interaction exists between the acid group of the stabilizer and the ester group on the monomer (methyl methacrylate). The H-bond anchors the stabilizer to the growing polymer particles, stabilizing the dispersion. Since there are no C–H bonds on the stabilizer, no hydrogen abstraction occurs and no stabilizer is incorporated into the final isolated polymer [44].

10.3.3 Polymerization in Fluorous Solvents

A fluorous biphasic system has been used to reduce the metal contamination arising in the copper-catalysed living radical polymerization of vinyl monomers.

Polymer Cracting i-Ionic Liquids

Acidic chloroaluminate ionic liquids are excellent media for polymer cracking reactions. With the huge quantities of polymers that need to be disposed of each year the ability to break them down into useful compounds for new synthesis or to use as liquid fuels is extremely important. While certain polymers such as poly(methyl methacrylate) are easily cracked into their constituent monomers that can be reused, the majority of polymers are extremely difficult to crack into useful organic compounds. However, merely dissolving polyethylene in acidic chloroaluminate ionic liquids containing a proton source results in the formation of a mixture of alkenes and cyclic alkenes [48]. The key compounds produced are shown in Figure 10.10.

The precise distribution of products that is obtained during the cracking process depends upon the temperature at which the reaction is carried out. Temperatures as low as 90°C are effective, which is considerably lower than those usually required, which are in the order 300–1000 °C. Another advantage of the chloroaluminate-based process is that aromatic compounds are only obtained in low concentrations compared to traditional methods that produce aromatic compounds in much higher yield.

Volatile alkanes   Low-volatile alkanes

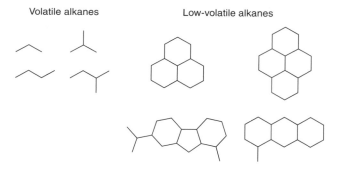

Figure 10.10

Copper(I) bromide and pentakis-N-(heptadecafluoroundecyl)-1,4,7-triazeheptane (**1** in Figure 10.9), along with an initiator, ethyl-2-bromoisobutyrate (**2** in Figure 10.9), in a perfluoromethylcyclohexane–toluene biphase efficiently catalyse the polymerization of methyl methacrylate, with a conversion of 76 % in 5 h at 90 °C. The resultant polymer has a $M_n = 11\,100$ and a molar mass distribution of 1.30. After polymerization, the reaction was cooled to ambient temperature, the organic layer removed and found to contain a copper level of 0.088 % (as opposed to 1.5 % if all the catalyst were to have remained in the polymer). A further toluene solution of monomer and **2** could be added, with efficient polymerization again occurring showing that reuseable catalysts for living polymerizations can be achieved using fluorous biphasic systems [46]. Suspension polymerizations have also been carried out in perfluorocarbon solvents [47].

10.4 CONCLUSIONS

Water is already widely used as a solvent for C–C coupling, metathesis and polymerization reactions. Supercritical carbon dioxide is being used industrially as an alternative to CFCs for polymerization reactions, showing that, if the application is suitable, there is no reason why these alternative solvents should not be used on an industrial scale. The use of $scCO_2$ may also lead to benefits in areas other than synthetic chemistry, for example in the processing of polymers. In just the last few years, these reactions have also been carried out using other alternative solvents in biphasic and multiphasic processes. The advantages offered by alternative solvents are important steps towards achieving current goals in C–C coupling, metathesis and polymerization reactions. For example, they can help increase the lifetime and reuse of the precious metal compounds used to catalyse these reactions, and lower the quantities of catalysts used. In addition to the replacement of volatile organic solvents, more benign reagents, such as replacing bromo- or iodoarenes with chloroarenes in Heck reactions, need to be exploited despite their lower reactivity. Again, alternative solvents could offer the answer. Clearly, a great deal more research needs to be carried out, but the studies described in this chapter indicate the vast potential that alternative solvents offer.

REFERENCES

1. Böhm V. P. W. and Herrmann W. A. *Chem. Eur. J.* 2000, **6**, 1017.
2. (a) Xu L., Chen W. and Xiao J. *Organometallics* 2000, **19**, 1123; (b) Mathews C. J., Smith P. J., Welton T., White A. J. P. and Williams D. J. *Organometallics* 2001, **20**, 3848.
3. (a) Hasan M., Kozhevnikov I. V., Siddiqui M. R. H., Steiner A. and Winterton N. *J. Chem. Res. Syrop.* 2000, 392; (b) Hasan M., Kozhevnikov I. V., Siddiqui M. R. H., Femoni C., Steiner A. and Winterton N. *Inorg. Chem.* 2001, **40**, 795.
4. Moineau J., Pozzi G., Quici S. and Sinou D. *Tetrahedron Lett.* 1999, **40**, 7683.
5. Nakamura Y., Takeuchi S., Zhang S., Okumura K. and Ohgo Y. *Tetrahedron Lett.* 2002, **43**, 3053.
6. Birdsall D. J., Hope E. G., Stuart A. M., Chem W., Hu Y. and Xiao J. *Tetrahedron Lett.* 2001, **42**, 8551.
7. Carroll A. B. and Holmes A. B. *Chem. Commun.* 1998, 1395.
8. Shezad N., Oakes R. S., Clifford A. A. and Rayner C. M. *Tetrahedron Lett.* 1999, **40**, 2221.
9. Adams C. J., Earle M. J., Roberts G. and Seddon K. R. *Chem. Commun.* 1998, 2097.
10. Rajagopal R., Jarikote D. V. and Srinivasan K. V. *Chem. Commun.* 2002, 616.
11. Moreno-Mañas M., Pleixats R. and Villarroya S. *Organometallics* 2001, **20**, 4524.
12. Chen D., Qing F.-L. and Huang Y. *Org. Lett.* 2002, **4**, 1003.

13. (a) Galante A., Lhoste P. and Sinou D. *Tetrahedron Lett.* 2001, **42**, 5425; (b) Barthelemy S., Schneider S. and Bannwarth W. *Tetrahedron Lett.* 2002, **43**, 807.
14. Le Boulaire V. and Grée R. *Chem. Commun.* 2000, 2195.
15. Fürstner A., Ackermann L., Beck K., Hori H., Koch D., Langemann K., Liebl M., Six C. and Leitner W. *J. Am. Chem. Soc.* 2001, **123**, 9000.
16. Csihony S., Fischmeister C., Bruneau C., Horváth I. T. and Dixneuf P. H. *New J. Chem.* 2002, **26**, 1667.
17. Grigg R. and York M. *Tetrahedron Lett.* 2000, **41**, 7255.
18. Kirkland T. A., Lynn D. M. and Grubbs R. H. *J. Org. Chem.* 1998, **63**, 9904.
19. Claverie J. P. and Soula R. *Prog. Polym. Sci.* 2003, **28**, 619.
20. Mecking S., Held A. and Bauers F. M. *Angew. Chem., Int. Ed. Engl.* 2002, **41**, 544.
21. Jiang Z. and Sen A. *Macromolecules* 1994, **27**, 7215.
22. (a) Claverie J. P. and Soula R. *Prog. Polym. Sci.* 2003, **28**, 619–; (b) Perrier S., Armes S. T., Wang X. S., Malet F. and Haddleton D. M. *J. Polym. Sci. Part A: Polym. Chem.* 2001, **39**, 1696.
23. Hoffmann U., Helmer-Metzmann F., Klapper M. and Müllen K. *Macromolecules* 1994, **27**, 3575.
24. Balakrishnan T. and Arivalagan K. *J. Polym. Sci. Part A: Polym. Chem.* 1994, **32**, 1909.
25. Scholsky K. *J. Supercrit. Fluids* 1993, **6**, 103.
26. Buback M. and Tups H. *Physica B and C* 1986, **139** and **140**, 626.
27. DeSimone J. M., Guan Z. and Elsbernd C. S. *Science* 1992, **257**, 945.
28. Guan Z., Elsbernd C. S. and DeSimone J. M. *Polym. Prep. (Am. Chem. Soc., Div. Polym. Chem.)* 1992, **34**, 329.
29. Combes J. R., Guan Z. and DeSimone J. M. *Macromolecules* 1994, **27**, 865.
30. (a) McCoy M. *Business* 1999, **77**, 11; (b) McCoy M., *Chem. Eng. News* 1999, June, 11; (c) Unattributed article, *Chemical Week*, 27 March 2002.
31. Van Bramer D. J., Shiflett M. B. and Yokozeki A. *US Pat.* 5345013, 1994.
32. Romack T. J., DeSimone J. M. and Treat T. A. *Macromolecules* 1995, **28**, 8429.
33. Canelas D. A. and DeSimone J. M. *Adv. Polym. Sci.* 1997, **133**, 103.
34. Romack T. J., Maury E. E. and DeSimone J. M. *Macromolecules* 1995, **28**, 912.
35. Beuerman S., Buback M., Isemer C. and Wahl A. *Macromol. Rapid Commun.* 1999, **20**, 26.
36. Mang S. A., Dokolas P. and Holmes A. B. *Org. Lett.* 1999, **1**, 125.
37. (a) Romack T. J., Kipp B. E. and DeSimone J. M. *Macromolecules* 1995, **28**, 8429; (b) Cooper A. I., Hems W. P. and Holmes A. B. *Macromol. Rapid Commun.* 1998, **19**, 353.
38. Hori H., Six C. and Leitner W. *Macromolecules* 1999, **32**, 2466.
39. Darensbourg D. J., Stafford N. W. and Katsurao T. *J. Mol. Catal. A: Chemical* 1995, **104**, L1.
40. Darensbourg D. J. and Holtcamp M. W. *Macromolecules* 1995, **28**, 7577.
41. Super M., Berluche E., Costello C. and Beckman E. J. *Macromolecules* 1997, **30**, 368.
42. Consani K. A. and Smith R. D. *J. Supercrit. Fluids* 1990, **3**, 51.
43. DeSimone J. M., Maury E. E., Menceloglu Y. Z., McClain J. B., Romack T. J. and Combes J. R. *Science* 1994, **265**, 356.
44. Howdle S. M. *Green Chem.* 2002, **4**, G29 and references therein.
45. Cooper A. I. and Holmes A. B. *Adv. Mater.* 1999, **11**, 1270.
46. Haddleton D. M., Jackson S. G. and Bon S. A. F. *J. Am. Chem. Soc.* 2000, **122**, 1542.
47. Mayes A. G. and Mosbach K. *Anal. Chem.* 1996, **68**, 3769.
48. Adams C. J., Earle M. J. and Seddon K. R. *Green Chem.* 2000, **2**, 21.

11 Alternative Reaction Media in Industrial Processes

The alternatives to volatile organic solvents that have been discussed in the preceding chapters offer the advantages of improved separation of products, and the isolation and reuse of expensive metal catalysts. Consequently, they have a great deal to offer the chemical industry. However, virtually all of the applications of alternative solvents in synthesis and catalysis described so far have only been conducted on a small scale in research laboratories, although there is precedent for the rapid development of new laboratory-scale biphasic technology into successful commercial processes if there are significant advantages over existing methods. The ultimate objective of all this research is to produce more efficient ways of conducting chemical reactions, thereby reducing pollution from volatile organic solvents, and making catalytic processes safer and more economically viable. In this chapter, we look at some examples where multiphasic and alternative solvent technologies are used on an industrial scale, and consider some of the reasons why other systems have not yet become commercial processes.

11.1 OBSTACLES AND OPPORTUNITIES FOR ALTERNATIVE MEDIA

Any new technology must overcome certain obstacles if it is to be applied successfully. These obstacles may be technical, financial, or may be due to lack of knowledge, or even the fear of being the first to put new science into practice. There is often a 'rush to be second' when faced with innovation, as many companies are reluctant to pioneer unproven technology because of the considerable financial risks involved [1]. Once the feasibility of the system has been demonstrated, there is frequently a flood of new applications. If initial obstacles can be overcome, the benefits of new solvent technology should provide many opportunities to improve process efficiency and reduce costs. Some of the barriers and benefits of using alternative solvents are shown in Figure 11.1. However, new technology requires investment in research, process development and equipment, and it may be many years before these costs can be recovered. It is therefore most likely that any commercialization of these systems will only occur when there

Chemistry in Alternative Reaction Media   D. Adams, P. Dyson and S. Tavener
© 2004 John Wiley & Sons, Ltd   ISBNs: 0-471-49848-3 (Cloth); 0-471-49849-1 (Paper)

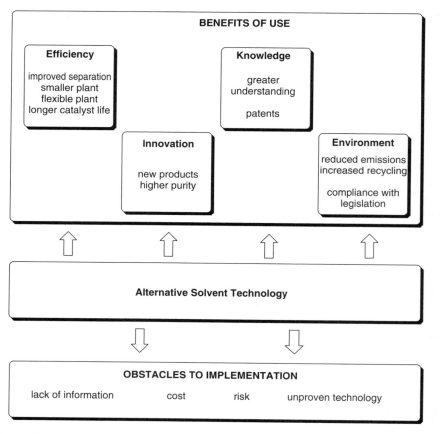

Figure 11.1 Benefits of use and obstacles to implementation for alternative reaction media in the chemical industry

is a major cost saving, a new or improved product is attainable, or if legislation prevents the use of an existing process.

As well as the financial costs of switching to alternative solvent technologies, a major barrier is a lack of fundamental physical data available for these solvents and their biphasic mixtures. Although more studies are reported every month, the information is dispersed in journals and technical reports, and frequently measurements are not conducted under comparable conditions which makes evaluation of data difficult. For example, knowledge of partition coefficients is of great importance when designing a biphasic system, but the sheer number of possible combinations of organic, aqueous, fluorous and ionic liquid solvents makes the acquisition of a comprehensive set of measurements a mammoth task. Chemical engineers also need to know other parameters such as heat capacity, flash point and viscosity if a safe process is to be designed.

Another barrier to implementation of biphasic technology is the question of how to choose a solvent system for a particular application. In general, a pair of solvents should be selected so that the product will be preferentially soluble in one phase, and the catalyst and substrate in the other. The partitioning of the catalyst into that phase should be as great as possible to ensure its effective retention and extended working lifetime. Secondary issues such as the phase behaviour on heating may be considered if a homogeneous system is required for the reaction. Broadly speaking, fluorous solvents and supercritical carbon dioxide (scCO$_2$) have low polarities, ionic liquids and water are polar, and organic solvents display a broad range of polarities. An ideal biphase should thus be attainable for any application, but the number of permutations makes the task of identifying the perfect combination difficult.

11.2 REACTOR CONSIDERATIONS FOR ALTERNATIVE MEDIA

In order to make the most of the properties of multiphasic systems, it may be necessary to redesign the entire process. They may not be considered as drop-in replacements for homogeneous reactions. However, any additional plant required at the reactor stage may be offset against the simplification of the separation. Distillation equipment is not normally needed when alternative media are employed. Ultrasound and microwave techniques have also been proposed as useful ways of promoting reactions between reactants in separate phases, through enhancement of mass transport by agitation of the interface and localized heating effects [2]. Supercritical fluids (SCFs) require specially constructed reactors because of the high pressures involved, and special pumps to achieve critical point.

11.2.1 Batch Reactors

Many chemical reactions are performed on a batch basis, in which a reactor is filled with solvents, substrates, catalysts and anything else required to make the reaction proceed, the reaction is then performed and finally the reactor is emptied and the resultant mixture separated (Figure 11.2). Conceptually, a batch reactor is similar to a scaled up version of a reaction in a round-bottomed flask, although obviously the engineering required to realize a large scale reaction is much more complicated. Batch reactors are suitable for homogeneous reactions, and also for multiphasic reactions provided that efficient mixing between the phases may be achieved so that the reaction occurs at a useful rate.

A large drawback of batch reactors is that a great deal of time is spent *not* performing chemical transformations: a batch process is inherently inefficient in terms of plant usage. Also, as the reaction proceeds and more product is formed, the composition of the batch changes, altering the physical properties of the reaction including viscosity, heat capacity and gas solubilities, all of which

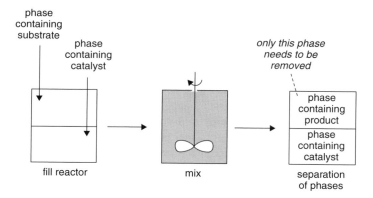

Figure 11.2   Schematic diagram of a biphasic process conducted in a batch reactor

may have undesirable effects on the process. Exothermic reactions may exhibit a further drawback because, as a batch reaction is scaled up, the ratio of volume to surface area becomes greater, and the efficiency of cooling and heat transfer is reduced. As the reaction heats up, its rate increases liberating yet more heat. In an extreme situation, a runaway reaction may occur. More usually, the reaction becomes less selective at higher temperatures, because alternative by-product forming pathways occur in addition to the desired reaction. The use of a biphasic system may actually help to control the effects of an exotherm. Because the rate of reaction, and therefore the amount of heat liberated, is controlled by mass transport between the two phases, the exotherm may be controlled simply by reducing the speed of stirring.

11.2.2   Flow Reactors

Heterogeneous reactions may also be conducted in flow reactors, in which re-agents are continuously added to the reactor, and products continually removed. The improved separation characteristics of heterogeneous reactions allow the possibility of leaving one of the phases in the reactor, and passing the other phase through it. This is common practice for heterogeneous gas–solid systems, such as catalytic hydrocarbon cracking, where the solid catalyst stays in the reactor and the gas is passed over it. Whilst this is solvent free, it is limited in application to volatile substrates, and cannot always be used as an alternative to liquid phase reactions. A biphasic mixture of liquids may be stirred or agitated on entry, and phase separation occurs after the mixture has left the reactor, as shown in Figure 11.3a. Although the composition of reactants varies gradually from one end of the reactor to the other, it will be consistent over time at any particular point, allowing the reactor to be designed accordingly. An alternative design suitable for biphasic liquid–liquid systems is a bubble reactor, in which the

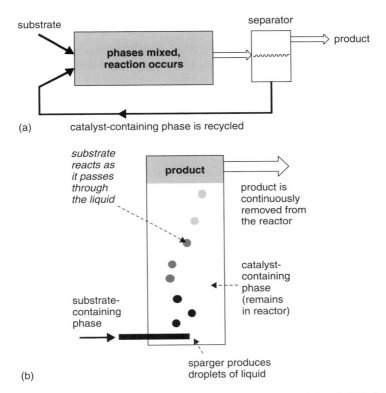

Figure 11.3 Schematic diagrams of (a) a continuous flow reactor and (b) a bubble flow reactor for biphasic reactions

liquid phase of lesser density is allowed to pass through the catalyst-containing phase (Figure 11.3b). The relatively high densities of ionic liquids and fluorous solvents should make them ideal for this kind of set-up.

Flow reactors generally have a much smaller reactor volume than batch reactors. For example, a 50 kg reaction that requires 2 h in a 100 litre reactor could be performed in less time in a flow reactor of only 1 litre volume, using a flow rate of 500 ml per minute. Flow reactors are ideal for SCFs because the lower volumes allow for more sensible reactor dimensions and wall thickness. Whereas scale up of a batch reactor requires building a larger plant, flow reactors may yield more product by increasing the speed through which the reaction mixture is passed through. Alternatively, parallel small-scale reactors may be built to increase throughput. Because reactor volumes are small, heat management is better and exotherms may be controlled more easily. More concentrated solutions may be used as less solvent is required for heat dissipation. Rapid heating and cooling may be achieved, allowing for faster reaction rates without unwanted by-products.

11.2.3   New Technology Suitable for Multiphasic Reactions

Ensuring a successful reaction between substances in separate phases may require unusual solutions to the problem of reactor design, and some ingenious methods of mixing have been developed. Biphasic reactions may be performed with membranes separating the two phases, and in this situation it is even possible for the two liquid phases to be miscible, as the membrane will keep them apart. A membrane reactor (shown in Figure 11.4) allows continuous reaction without any mixing of the two phases [3]. If required, the membrane may be coated in biphilic phase transfer catalyst (PTC) functional groups to enhance reactivity, or else a PTC added to increase the transfer of reactants through the membrane [4].

The two phases enter the reactor by different pipes, the reaction takes place across the membrane and the product and waste reagent leave via separate pipes. This set-up has been applied to the oxidation of benzyl alcohol to benzaldehyde using sodium hypochlorite as oxidant [5]. In this case the organic compounds cannot pass through the membrane so the aqueous phase remains free of organic contamination, and the inorganic oxidant may be regenerated electrochemically and recycled, giving a reactor set-up that produces almost no waste.

The use of a rotating reactor to provide efficient mixing has been proposed and tested for viscous polymerizations and acid-catalysed rearrangement reactions [6]. The principle of these spinning disc reactors is that the centrifugal force

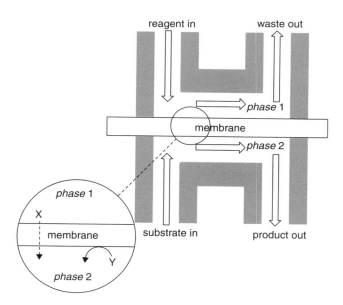

Figure 11.4   Schematic diagram of membrane reactor. The reaction is designed so that reagent X may pass through the membrane and is therefore available for reaction. The substrate and product cannot pass through and the phases remain separate

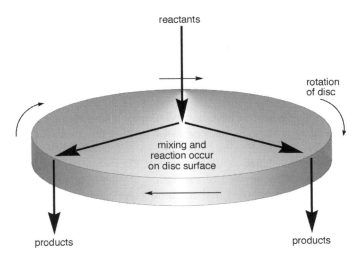

Figure 11.5   Schematic diagram of a spinning disc reactor. Products are collected from a cooled trough positioned below the disc

created on the surface of the disc produces thin films that are subject to high shear forces and thus mixed efficiently. The disc is grooved to further improve mixing. A schematic diagram of the disc reactor is shown in Figure 11.5. The rotation also drives the reaction mixture rapidly towards the edge of the disc. Residence time on the disc is therefore very short, typically a few seconds, and this allows the use of higher temperatures without adverse by-product formation. Additionally, a solid catalyst may be glued to the disc if required. These systems have not yet been used to accelerate liquid–liquid biphasic reactions, but the enhanced mixing provided by these systems is clearly applicable to multiphase chemistry.

11.3   INDUSTRIAL APPLICATIONS OF ALTERNATIVE SOLVENT SYSTEMS

The abundance of water, combined with it being nontoxic, has been partly responsible for its deployment in an important biphasic industrial process. Phase transfer catalysis is now well established as a useful method for large-scale synthesis, primarily because it allows the use of low cost inorganic reagents. $ScCO_2$ is also inexpensive and has found several industrial applications, and the engineering technology necessary is well understood through its use as an extraction fluid for decaffeination and other processes, although its use as a solvent for synthetic chemistry on a large scale is only just becoming a reality. The cost of ionic liquids and fluorous solvents is comparatively very high and their toxicity remains unknown, which could potentially restrict their applications. However, in processes where these solvents are used to immobilize the catalyst then, once in place, the solvent should last for a considerable time.

11.3.1   The Development of the First Aqueous–Organic Biphasic Hydroformylation Plant

The synthesis of aldehydes via hydroformylation of alkenes is an important industrial process used to produce in the region of 6 million tonnes a year of aldehydes. These compounds are used as intermediates in the manufacture of plasticizers, soaps, detergents and pharmaceutical products [7]. While the majority of aldehydes prepared from alkene hydroformylation are done so in organic solvents, some research in 1975 showed that rhodium complexes with sulfonated phosphine ligands immobilized in water were able to hydroformylate propene with virtually complete retention of rhodium in the aqueous phase [8]. Since catalyst loss is a major problem in the production of bulk chemicals of this nature, the process was scaled up, culminating in the Ruhrchemie–Rhône-Poulenc process for hydroformylation of propene, initially on a 120 000 tonne per year scale [9]. The development of this biphasic process represents one of the major transitions since the discovery of the hydroformylation reaction. The key transitions in this field include [10]:

1   Development of a heterogeneously catalysed processes based on cobalt.
2   Introduction of a high pressure process employing homogeneous cobalt catalysts.
3   Replacement of cobalt with rhodium.
4   Tuning the catalyst by modifying the ligands.
5   Introduction of the aqueous biphase process.

What is particularly remarkable about the development of the biphasic process is the speed in which it was developed following the initial discovery. Over the next few pages some of the key steps in this process will be described [11].

The development of the first water-soluble rhodium hydroformylation catalyst in 1975 was directly triggered by the first preparation of TPPTS (see Figure 11.6), the water-soluble phosphine used today in conjunction with the rhodium catalyst. TPPTS was originally prepared by sulfonation of triphenylphosphine using oleum, and the synthesis of this important ligand has subsequently been refined many times [12]. Experiments combining TPPTS with rhodium salts followed by an evaluation of the activity in various catalysed reactions in water showed them not only to be very effective catalysts, but also retention of the rhodium in the aqueous phase was extremely high. Rhône-Poulenc filed a series of patent applications and since the hydroformylation reaction shown in Scheme 11.1 showed considerable promise the company began to develop this further.

A few years later, Ruhrchemie joined forces with Rhône-Poulenc to develop a continuous biphasic hydroformylation process since Rhône-Poulenc had no

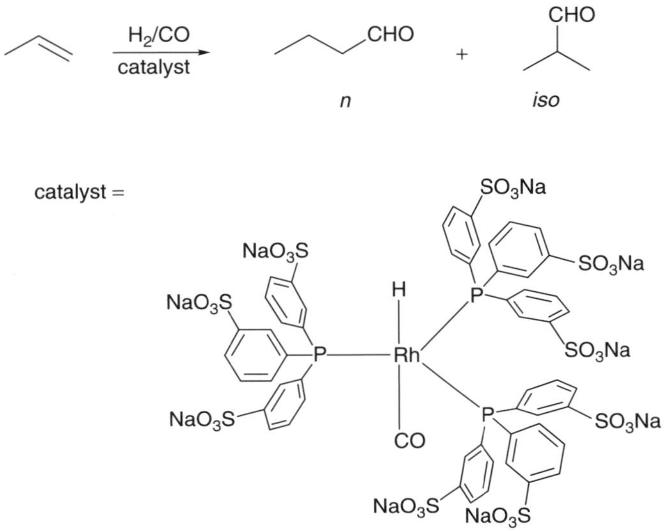

Scheme 11.1

experience in hydroformylation. While many benefits were envisaged, virtually everything to do with the technology was unknown. In less than 2 years a completely new process that had no precedent was developed and tested and, using a scale-up factor of 1 : 24 000, the first production went into operation. The plant opened in 1984 and had an initial capacity of 120 000 tonnes per year. The beauty of the process is that the substrates, propene, H_2 and CO, are gases at room temperature and dissolve in the water, whereas the product, which contains less than 4 % iso-butyraldehyde, is a liquid under the conditions of the process and forms a second phase which is easily removed and is virtually free from rhodium contamination. In addition, during the production of the first 2 million tonnes of n-butyraldehyde only 2 kg of rhodium were lost, which is in the parts-per-billion range.

Since the plant went online, a great deal of fundamental research has been conducted and the process has been scaled-up further and new plants have come into operation. One of the main things to have been investigated further is the catalyst. The homogeneous rhodium catalyst can easily be tuned by modifying the ligands surrounding it. With this in mind, a number of other water-soluble phosphines have been examined [13]. Figure 11.6 shows how higher activities can be obtained for the hydroformylation of propene by replacing the TPPTS with other ligands.

Clearly, other phosphine ligands have a pronounced effect on the activity at the rhodium centre. Not only does the activity and percentage of the n-product increase, but the amount of phosphine that is required relative to the rhodium also

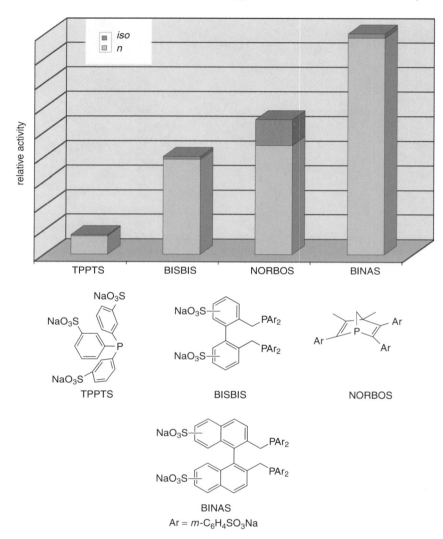

Figure 11.6 Effect of varying the water soluble phosphine ligand on the hydroformylation of propene to butyraldehyde. P:Rh ratio = TPPTS 80; BISBIS 7; NORBOS 14; BINAS 7

decreases. Despite these apparent advantages, the phosphine ligand that continues to be used in the Ruhrchemie–Rhône-Poulenc process is TPPTS, a decision undoubtedly based on lower cost and ease of preparation of this ligand.

The economic and environmental benefits of the Ruhrchemie–Rhône-Poulenc process have been closely scrutinized since the plant has been in operation.

Overall the cost of the production is reduced, but it is the benefits to the environment that are the greatest [14]. These benefits can be summarized as:

1 Use of water in place of toxic organic solvents.
2 Conservation of energy resources as the process operates under milder conditions compared to conventional methods.
3 Higher selectivity towards the desired *n*-butyraldehyde product.
4 Very low loss of the toxic rhodium catalyst.

Overall, the efficiency is extremely impressive and the environmental emissions are almost zero.

11.3.2 Other Examples of Processes Using Water as a Solvent

The Ruhrchemie–Rhône-Poulenc hydroformylation process is certainly the most famous large-scale biphasic operation. However, a number of other processes have been scaled-up and are employed industrially. Rhône-Poulenc produces a number of vitamin precursors using an aqueous-organic C–C coupling reaction which uses a rhodium–TPPTS catalyst [15]. This process operates on a relatively small scale. Another process that uses the Suzuki reaction has also been commercialized and is illustrated in Scheme 11.2 [16]. Apart from the environmental benefits associated with a biphasic process, the reaction is particularly attractive as it uses a chloroaromatic as the starting material in place of the bromo or iodo analogues normally required in C–C coupling reactions. From an ecological perspective, the use of chloro compounds is considerably preferable to the alternatives.

11.3.3 Scale-Up of PTC Systems

Phase transfer catalysis has been widely employed in commercial synthetic applications, primarily because of its ability to replace organic amine and alkoxides bases with low cost metal hydroxides. An example of this from PTC Organics

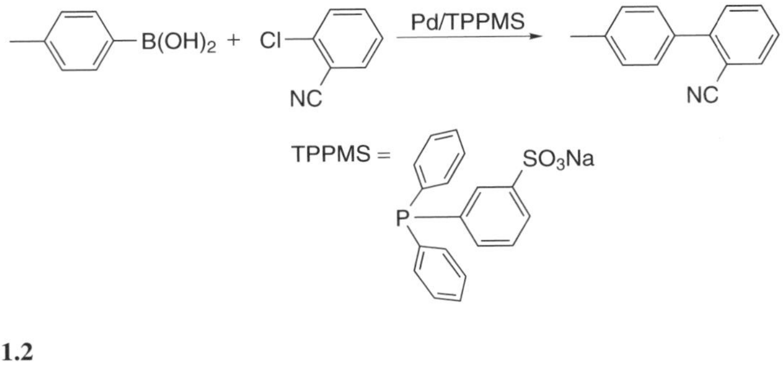

Scheme 11.2

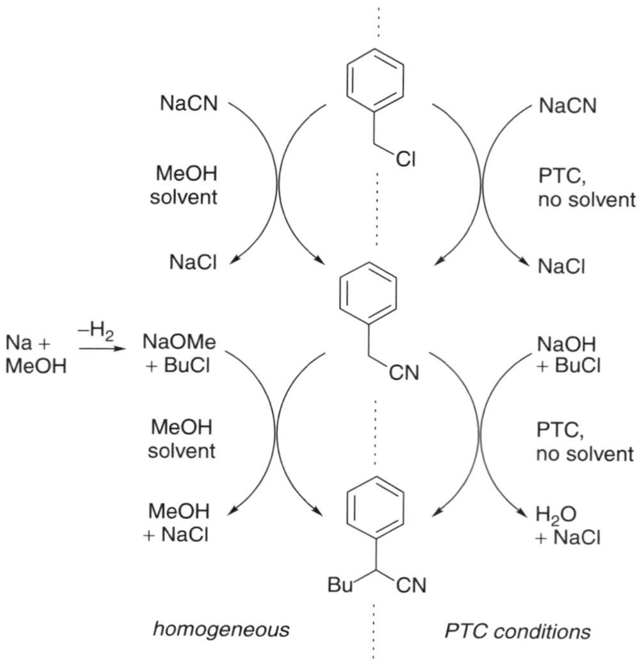

Scheme 11.3

Inc. involves the replacement of sodium methoxide with sodium hydroxide in a two-step transformation of benzyl chloride to 2-phenylhexanenitrile, as shown in Scheme 11.3 [17]. In the cyanation step, replacing the methanol solvent with a PTC reduces the degree of solvation via H-bonding around the cyanide anion, and thus improves its activity as a nucleophile. The outcome is that the homogeneous method requires a reaction time of 11 h, whereas the PTC method only requires 4 h. Another advantage is that it is no longer necessary to add a scrubber or chiller to recover the volatile methanol at the reactor outlet. In the second step of the reaction (alkylation), use of NaOH and a PTC eliminates the need for NaOMe, which is usually formed by reaction of sodium metal with methanol. This process liberates hydrogen and is clearly hazardous on a large scale. Other benefits to the overall process by using the PTC route include an increase in yield from 75 to 85 %, a reduction in process time from 18 to 10 h and an increase in plant capacity of 105 %. The drawbacks associated with the PTC method were a slight increase in the quantity of waste water produced, and the need for additional toluene to be added at a final distillation stage to ensure efficient separation of product from catalyst. Even accounting for the additional toluene, an overall reduction of organic waste of 43 % was achieved.

Phase transfer catalysis exemplifies how a laboratory curiosity can successfully make the transition to becoming a routine process, and PTC is now commonly featured in many general chemistry texts. It has been applied to numerous processes, both large and small scale [18].

11.3.4 Thomas Swan Supercritical Fluid Plant

Supercritical fluids have been used on an industrial scale for many decades, notably for the polymerization of ethene and for extraction of caffeine from coffee beans as described in Chapter 6. However, it is only in the past few years that SCFs have been taken seriously as reaction media for chemical synthesis on a large scale. Perhaps the major obstacle has been the manufacture of reactors capable of withstanding the high pressures required to reach critical point. Because of the high pressures, the wall thickness of reactors must increase rapidly as the reaction volume is increased, leading to huge plants that would be extremely expensive to construct. This makes the construction of large batch reactors for SCF applications unfeasible. It is therefore not surprising that the first supercritical plant constructed for chemical synthesis is a flow reactor, which reduces both the cost and risk associated with running large volume reactions at high pressure. If a higher volume of product is required, the flow reactor is simply run for a longer time, or parallel reactors may be built. Flow reactors also allow better control over reaction conditions in SCFs: in a batch reactor the composition of the system, and therefore the critical point, changes as the reaction proceeds and more substrate is converted to product. By contrast, in a flow reactor the composition at any point in the reactor should stay constant with time. Recently, the first continuous flow SCF plant, with a capacity of up to 1000 tonnes per year, has been developed in the UK by Thomas Swan & Co. [19]. Since the SCF systems combine the mass transport properties of gases with the mass density of liquids, increased reaction kinetics leads to high throughputs from a relatively small reactor system. In addition, by using a continuous flow reactor where the substrates are only exposed to reaction conditions for a short time, it becomes possible to have reactions that are controlled kinetically, as opposed to thermodynamically, affording products which would be difficult to obtain under conventional conditions. The plant has been designed to be versatile, and can, in theory, be used to perform any reaction that proceeds in $scCO_2$. It is envisaged that the plant will be used for contract synthesis as required. The initial manufacturing process makes use of the infinite solubility of hydrogen in $scCO_2$: it is a hydrogenation reaction using a Pd catalyst, the formation of trimethylcyclohexanone from isophorone [20]. The supercritical process in this case gives greatly improved selectivity and avoids the over-hydrogenation that is usually seen in conventional solvents, and therefore gives a much purer product as shown in Scheme 11.4. The

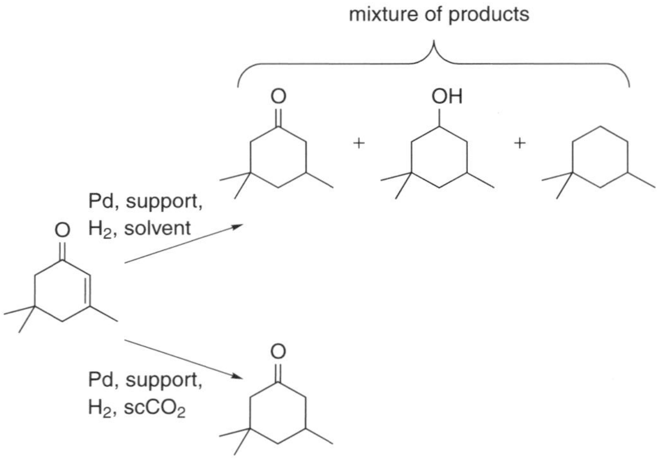

Scheme 11.4

scaled up scCO$_2$ process gives quantitative conversion and requires only 1.2 g of a heterogeneous catalyst containing 2 % Pd, per kg of product. The real advantage of using SCFs for this process is that the product is obtained free of impurities, and the SCF solvent is completely removed by depressurization: there is no need for any purification step. The process is therefore extremely efficient, and other hydrogenations, and Friedel–Crafts reactions are planned in the near future.

11.3.5 Other Applications of Supercritical Carbon Dioxide

Inhale Therapeutic Systems has recently spent US\$ 63 million on the acquisition of Bradford Particle Design, a UK company which has developed the use of SCF technology for the close control of particle size, particularly for the manufacture of pharmaceuticals [21]. The current plant is capable of producing up to 20 kg of product per day. In this application, scCO$_2$ is used as an antisolvent for the precipitation of micron-sized crystallites from solution. The scCO$_2$ simultaneously extracts the solvent and disperses and mixes the solution, which leads to the formation of dry, solvent free particles with a very small size distribution. By controlling the pressure and temperature of the SCF, it is possible to control the particle size and even select a particular polymorphic form. For example, the anti-asthmatic drug salmeterol xinafoate (Figure 11.7) exists in two polymorphic forms, neither of which are readily obtained in a pure form by crystallization using conventional solvents. Polymorph I (Mp 124 °C) is usually formed contaminated with a small amount of polymorph II (Mp 136 °C). Using scCO$_2$ to disperse and crystallize the drug from a solution of the drug in acetone and methanol, either polymorph may be obtained in a pure form, with the

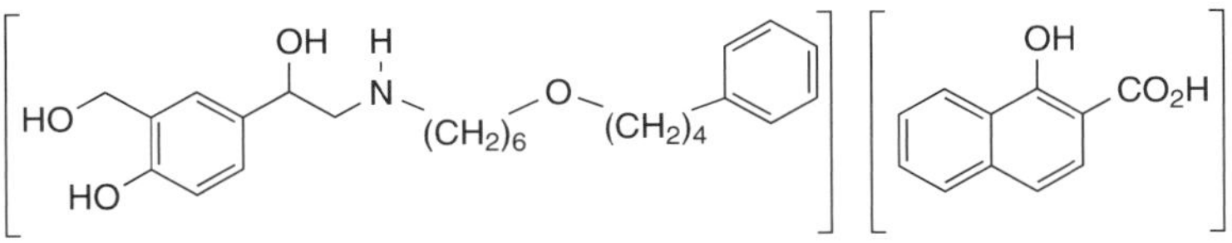

Figure 11.7 Salmeterol Xinafoate

Table 11.1 The effect of crystallization conditions on the polymorphic form of salmeterol xinafoate

Conditions	Result
Liquid solvents	Mixture of I and II
scCO$_2$, 100 bar, 60 °C	Mixture of I and II
scCO$_2$, 250 bar, 40 °C	I, pure
scCO$_2$, 250 bar, 90 °C	II, pure

selection of the polymorph being achieved simply by changing the temperature of the system (Table 11.1) [22]. Different polymorphs of pharmaceuticals may have different physiological properties and may be patented separately from one another, so this selectivity is of considerable value.

Because small particles can be produced in SCFs without the need for grinding or milling, it is possible to produce crystallites with undamaged surfaces, which have improved aerodynamic qualities for certain applications. There is a great deal of interest in the development of drug formulations that can be delivered using inhaler technology, and the size and shape of the particle is critical if the active ingredient is to be delivered efficiently. The use of SCF technology is greatly expanding the range of drugs that may be delivered via inhalers, and recent examples include inhaled insulin for the treatment of diabetes, and dronabinol (a synthetic tetrahydrocannibinol) for the treatment of weight loss in AIDS patients and the nausea induced as a side-effect of chemotherapy. Earlier generations of inhaler used CFCs as propellants, but compressed air may be used if the aerodynamic properties of the crystallites are suitable, and so this may perhaps be viewed as a case where scCO$_2$ is replacing CFCs, albeit indirectly.

As described in Chapter 10, scCO$_2$ has replaced the use of CFCs as a medium for polymerization of Teflon and other fluoropolymers, and DuPont have recently built a US$ 275 million plant capable of making 1000 tonnes of polymer per annum. The plant uses CO$_2$ technology in a process that generates less waste during manufacture, and produces a grade of polymer that the manufacturer claims has enhanced performance and processing capabilities [23]. The use of scCO$_2$ is now also being developed by Praxair and Supercritical Systems Inc. for the removal of photoresist materials from semiconductor wafers during the manufacture of microprocessor chips [24]. This is usually performed using a plasma process to degrade the photoresist, followed by removal with VOCs, but the supercritical process offers definite advantages over the conventional approach. In particular, SCFs have effectively no surface tension, which allows them to penetrate the microscopic channels and trenches that have been etched on the wafers ensuring that the cleaning process is efficient and reliable. The features on the wafers are smaller than 0.2 μm and it is imperative that all of the photoresist is removed. A small amount of co-solvent is used, and it is thought that the SCF helps transport the cleaning solvent to the sub-micron structures. Whilst this is not a synthetic use of scCO$_2$, it is indicative of the advantages that supercritical technology can bring, and the trend towards consideration of SCFs when designing new processes.

11.4 OUTLOOK FOR FLUOROUS SOLVENTS AND IONIC LIQUIDS

Processes using fluorous solvents and ionic liquids as solvents have not yet been commercialized, primarily because of cost. It is likely that they will find their first applications in small scale, high value products such as pharmaceuticals. The ability of biphasic systems to improve separation and reduce the quantity of metal catalyst present in the final product, which is vitally important in the case of medical products where even traces of toxic metals will prevent a potential drug from being licensed. Most applications will be where a product of superior form or purity is obtained. Industrial use of these solvents will require efficient partitioning of the catalyst into the fluorous or ionic liquid solvent to ensure that the catalyst has an extremely long lifetime and that product contamination is kept to a minimum. In addition, the product should be partitioned out of the catalyst phase, otherwise an additional solvent extraction step, possibly using a VOC, may be required to isolate the product. This would require further reactor complexity, and would negate some of the advantages of using the biphasic system in lowering the quantities of VOCs employed in the process.

One large scale application of an ionic liquid is as a base to remove HCl during the preparation of alkoxyphenylphosphines. BASF use *N*-methylimidazole which forms the ionic liquid [Hmim]Cl and forms a second liquid layer that may easily be removed from the system and recycled via phase separation. Although the ionic liquid is not employed as a solvent, the process makes use of biphasic technology to give a more efficient process which avoids an inconvenient filtration step.

The ideal systems for these media are those which do not require any additional solvent, and in which the substrate is more soluble than the product, leading to preferential rejection of the product from the catalyst phase. For fluorous reactions, this would include oxidation reactions where oxygenated products are typically more polar than the substrates. In ionic liquids it is products less polar than the substrates that will normally be less soluble, although the ability to tune the structure of ionic liquids to match a particular application must not be forgotten. For instance, a H-bonding group on the ionic liquid would ensure preferential solubility of an alcohol substrate over an alkene product in a dehydration–elimination reaction.

In the case of fluorous systems, it is the need to produce fluorous ligands that is currently economically prohibitive. For ionic liquids, it is the manufacture of the solvents themselves, which are very different from organic solvents and extremely difficult to obtain in a pure form because distillation methods cannot be used. However, it is expected that the principles of supply and demand will lead to large drops in prices for these chemicals in the immediate future. Until now the market has been limited to research scale quantities. However, scale up of production will lead to lower prices, more applications and greater demand. Certain imidazolium and pyridinium based ionic liquids are now available commercially, although only in gram quantities and at a cost equivalent to approximately €4000 (US$4000) per kg [25][1]. Scaled-up production by other manufacturers and lower prices are expected this year. Despite the current high prices, the listing of these ionic liquids in chemical suppliers catalogues is indicative of the fact that these solvents are now being taken seriously as commercial products, and any further breakthroughs in synthetic methodology will also lead to decreases in cost. It would be premature to predict the future based on current prices.

Fluorous and ionic liquids are often described as 'environmentally friendly', 'clean' or 'green' solvents, but these claims must be viewed with a degree of caution. Whilst they evidently have the potential to reduce emissions of VOCs in a carefully designed multiphasic process, these benefits must be offset against chemical and energy requirements during the synthesis of the fluorous and ionic solvents themselves. In addition, there are genuine concerns over the extremely long lifetime of perfluorinated compounds in the atmosphere, and a manufacturer wishing to scale up a fluorous process would have to convince regulators that emissions of the solvent could be kept down to an extremely low level. Ionic liquids are nonvolatile, and so present less of a problem, although they can still find there way into the environment through spillage. However, at present little is know about their toxicity and fate in the environment, and more study is required in this area.

11.5 CONCLUSIONS

By publishing the first descriptions of biphasic reactions, the pioneers of alternative reaction media have caused major changes in the way we think about chemical reactions. It is no longer instinctive to assume that everything must be dissolved together in a single phase for a reaction to occur effectively, and many chemists are now aware of the advantages to be gained in terms of separation and efficiency that may be achieved using multiphasic reactions. We have seen that there is no limit to the range of reactions that may be performed in these

---

[1] Value estimated from a 25 g batch of [bmim][PF$_6$] at £65GBP, using £1 $\approx$ €1.50 $\approx$ US$1.50.

alternative media. All the major types of reaction have now been successfully tested under some or all of the systems described in this book; the list of studied reactions grows all the time, and certain reactions have been found that only take place in alternative solvents. Whilst aqueous systems and $scCO_2$ have already found applications on an industrial scale, fluorous and ionic liquid processes have yet to be commercialized. For engineering and economic reasons, however, it is unlikely that these will be used simply as drop-in replacements for existing processes: there must be an added advantage such as ease of separation, improved yield, a route to new chemical products unattainable by other methods, or the use of lower cost substrates. Catalysts may be used and recycled effectively via immobilization in water, ionic liquids and fluorous solvents, and the solvents themselves may frequently be recycled. It is most likely that the fluorous and ionic liquids systems will find their first scaled-up applications in the production of high value pharmaceuticals, where the cost of the reaction medium may be offset by the increased purity that a biphasic reaction can offer, or where an important target molecule cannot be obtained under conventional conditions. Even if fluorous and ionic liquids never mature to become large commercial processes, then the concepts of multiphase catalysis and the science discovered in their investigation will certainly be used. The biphasic approach is by no means restricted to the fluorous, ionic liquid and aqueous systems described in this book. For example, cyclohexane and acetonitrile will form a biphase at room temperature. Of course this does not help to eliminate the use of volatile organic solvents, but may still provide improved separation in a carefully designed biphasic reaction.

So what is the future for these technologies? Should we expect the chemical industry to follow the lead of other high-technology industries such as the electronics and computing industries which are constantly moving towards products which give increased utility in a smaller package? Although this is already happening to some extent, with lower dosage medicines and detergents the subject of current investigation, many chemical products will always be required in large volumes. As well as reduction in the volume of product needed to achieve the same effect, there is also a move towards the design and construction of smaller chemical plants and processes that are capable of an increase in throughput and efficiency. The concept of process intensification involves a move away from large-scale batch reactions and their associated inefficiencies, towards flow reactors in which the actual volume in which the reaction occurs is very small indeed. The idea of microflow reactors in which the reacting volume is just a few microlitres has been the subject of much speculation [26]. In a batch process, the reactor must be loaded, started, and then stopped and emptied after the reaction is complete – it may only be in use for chemical transformation for a fraction of the time, and must be large enough to hold all the reactants and associated solvents. By contrast, a continuous flow reactor may be working 100 % of the time, bar maintenance, and can be many times smaller than a batch reactor. Flow reactors work best with heterogeneous reactions, and this is a real opportunity for innovation using alternative reaction media.

REFERENCES

1. Ritter S. K. *Chem. Eng. News* 2002, **80**(47), 19.
2. Thompson L. H. and Doraiswamy L. K. *Ind. Eng. Chem. Res.* 1999, **38**, 1215; (b) Mason T. J. and Cintas P. In *Handbook of Green Chemistry and Technology*, Clark J. H. and Macquarrie D. J. (eds), Blackwell, Oxford, 2002, p. 372; (c) Varma R. S. *Green Chem.* 1999, **1**, 43.
3. (a) Stanley T. J. and Quinn J. A. *Chem. Eng. Sci.* 1987, **42**, 2313; (b) Jachuck R. In *Handbook of Green Chemistry*, Clark J. H. and Macquarrie D. J. (eds), Blackwell, Oxford, 2002, p. 366.
4. Okahat Y., Ariga K. and Seki T. *Chem. Commun.* 1985, 921.
5. (a) Grigoropoulou G., Clark J. H., Hall D. W. and Scott K. *Chem. Commun.* 2001, 547; (b) Hall D. W., Grigoropoulou G., Clark J. H., Scott K. and Jachuck R. J. *Green Chem.* 2002, **4**, 459.
6. (a) Boodhoo K. V. K. and Jachuck R. J. *Green Chem.* 2002, **2**, 235; (b) Boodhoo K. V. K. and Jachuck R. J. *Appl. Thermal. Eng.* 2000, **20**, 1127.
7. Whyman R. *Applied Organometallic Chemistry and Catalysis*, Oxford University Press, Oxford, 2001.
8. Kuntz E. Fr. Pat. 2314910, 1975.
9. Cornils B. and Kuntz E. G. *J. Organomet. Chem.* 1995, **502**, 177.
10. Cornils B. *J. Mol. Catal. A: Chemical* 1999, **143**, 1.
11. Cornils B. *Org. Process Res. Dev.* 1998, **2**, 121.
12. Herrmann W. A., Albanese G. P., Manetsberger R. B., Lappe P. and Bahrmann H. *Angew. Chem. Int. Ed. Engl.* 1995, **34**, 811.
13. Bahrmann H., Bach H., Frohning C. D., Kleiner H. J., Lappe P., Peters D., Regnat D. and Herrmann W. A. *J. Mol. Catal. A: Chemical* 1997, **116**, 49.
14. Cornils B. and Wiebus E. In *Aqueous-Phase Organometallic Catalysis*, Cornils B. and Herrmann W. A. (eds), Wiley-VCH, Weinheim, 1998, pp. 259–268.
15. Mercier C. and Chabardes P. *Pure Appl. Chem.* 1994, **66**, 1509.
16. Haber S. and Kleinert H. J. DE Appl. 19527118 and 19535528, 1997.
17. Halpern M. *PTC Commun.* 1996, **2**, 1.
18. (a) Sharma M. In *Handbook of Phase Transfer Catalysis*, Sasson Y. and Neumann R. (eds), Chapman & Hall, London, 1997, p. 168; (b) Boswell C., *Chemical Market Reporter*, 4 November 2002.
19. (a) Institute of Applied Catalysis *iAc News* 2002, **9**, 1; (b) Press release PA66/02, University of Nottingham, 11 July 2002; (c) Gray W. K., Smail F. R., Hitzler M. G., Ross S. K. and Poliakoff M. *J. Am. Chem. Soc.* 1999, **121**, 10711.
20. Ross S. K., Smail F. R., Sellin M., Amandi R., Licence P. and Poliakoff M. Paper presented at 4th *International Symposium on High Pressure Process Technology and Chemical Engineering*, September 22–25, 2002, Venice, Italy.
21. (a) *Business Wire* 2 May 2002; (b) Bradford Particle Design, www.bdp.co.uk.
22. Shekunov B. Y., Edwards A. D., York P. and Cranswick L. M. D. *The Synchrotron Radiation Source Scientific Reports: Chemical Crystallography* 1997–1998, 67.
23. (a) McCoy M. *Business* 1999, **77**, 11; (b) McCoy M. *Chem. Eng. News* 1999, 11; (c) Unattributed article. *Chemical Week* 27 March 2002.
24. (a) Unattributed article. *Business Wire*, 28 September 2000; (b) www.cleanwafer.com.
25. Acros Organics, Geel, Belgium, www.acros.com.
26. (a) Haswell S. J., Middleton R. J., O'Sullivan B., Skelton V., Watts P. and Styring P. *Chem. Commun.* 2001, 391; (b) Fletcher P. D. I., Haswell S. J., Pombo-Villar E., Warrington B. H., Watts P., Wong S. Y. F. and Zhang X. *Tetrahedron* 2002, **58**, 4735.

Index

Page references followed by 't' refer to tables, those followed by 'f' refer to figures.

acceptor number (AN)   17t
acetic acid   99
acetone   13t, 17t, 19f, 20t, 22t, 27t, 51t, 85t
   as solvent   23, 79, 83
acetonitrile   11t, 13t, 17t, 19f, 20t, 22t, 51t, 85t, 152f
   biphase formation by   234
   enantioselectivity   102
   as reaction medium   197
   as solvent   23
acetophenone   187
acetylacetonate   211f
acrylates   210
acrylonitrile   208
acyl chloride   200
acylation   200
acylium   200
adipic acid   172–173f, 181, 187
alcohols   4t, 5t
   aliphatic   188
   as by-product   184
   fluorinated   143
   H-bonding by   13
   hydrogenation of   170
   in oxidization   181–182, 189
   as protic solvent   23
   reaction with propriolates   68
   solubility induction by   106
   as substrate   233
aldehydes
   in condensation reactions   181–182
   in hydroformylation   171–172, 224
   in hydrogenation   170

in oxidation   185, 188–189, 191
   reaction with phosphites   176
alkanes   181–192
   elimination of   161
   fluorinated   57–58
alkenes
   aliphatic   186
   in alkoxycarbonylation   75
   cyclic   214
   dichlorocyclopropanation of   119
   in Diels–Alder reactions   149
   in hydroformylation   67, 224
   in hydrogenation   75, 91, 161–171
   long chain   174
   in metathesis   145, 195, 202–203
   in oxidization   181–182, 184–188
   in phosphine formation   61, 200
   polymerization of   206
   solubility of   172, 233
1-alkenes   169
alkoxides   112, 145, 227
alkoxycarbonylation   75
alkoxyphenylphosphines   232
alkyl acrylates   178
alkyl iodide   62
1-alkyl-3-methylimidazolium   82
alkylation   228
alkyldimethylammonium salts
   polystyrene bound   125
alkylimidazoles   80
1,3-alkylmethylimidazolium triflate   80
alkylpyridinium chlorides   75
alkyls   145, 160–161, 165
   chains in   167

Chemistry in Alternative Reaction Media   D. Adams, P. Dyson and S. Tavener
© 2004 John Wiley & Sons, Ltd   ISBNs: 0-471-49848-3 (Cloth); 0-471-49849-1 (Paper)